INTRODUCTION TO MODERN POWER ELECTRONICS

Edited by
Galang, Maureen R.

Delve Publishing

Introduction to Modern Power Electronics

Edited by Galang, Maureen R.

Edition 2018

Delve Publishing

Published by
Delve Publishing LLC
708 3rd Avenue, 6th Floor
New York
NY 10801
United States of America
www.delvepublishing.com

ISBN: 978-1-68095-906-2 (Hardback)

Printed in United States of America

Library of Congress Control Number: 2016920465

About the Editor

Galang, Maureen R.

Maureen Reyes Galang received her Bachelor's degree in Electrical Engineering from Technological University of the Philippines, Manila in 2010. During her first career, Engr. Galang has been professionally involved in producing high quality 3D models, i.e. electrical equipment and devices for automotive and other machineries. Later of her profession focuses on building construction, i.e. residential, commercial and industrial. She became a highly skilled Electrical Engr. in Quantity Surveying and Cost Analysis bringing cost wise bids for clients.

Table of Contents

3 DC TO DC CHOPPERS

4 DC-TO-DC CONVERTERS

5 DC TO AC CONVERTORS

6 AC TO AC CONVERTORS

8 SWITCHED-MODE POWER SUPPLY

List of Abbreviations

AC	Alternating Current
APOD	Alternative Phase Opposition Disposition
ASD	Adjustable Speed Drives
BJT	Bipolar Junction Transistor
BOM	Bill of Materials
BOS	Balance of System
CMOS	Complementary Metal–Oxide–Semiconductor
CRT	Cathode Ray Tubes
CSI	Current Source Inverters
CVT	Constant Voltage Transformers
DRUPS	Diesel Rotary Uninterruptible Power Supply
EMF	Electromotive Force
EMI	Electromagnetic Interference
FET	Field-Effect Transistor
GTO	Gate Turn-Off Thyristors
HBT	Heterojunction Bipolar Transistor
HDMC	Hybrid Direct Matrix Converter
HEV	Hybrid Electric Vehicles
HVDC	High-Voltage Direct Current
IC	Integrated Circuits
ICE	Internal Combustion Engine
IGBT	Insulated-Gate Bipolar Transistor

LDO	Low-Dropout Regulators
MOS	Metal–Oxide–Semiconductor
PD	Phase Disposition
PDM	Pulse-Duration Modulation
PFC	Power Factor Correction
PMT	Photo Multiplier Tubes
POD	Phase Opposition Disposition
PWM	Pulse-Width Modulation
RCC	Ringing Choke Converter
SCR	Silicon Controlled Rectifier
SHE	Selective Harmonic Elimination
SMPS	Switch-Mode Power Supply
SOI	Silicon-On-Insulator
SPWM	Sinusoidal Pulse-Width Modulation
SPWM	Synchronized-Pulse-Width-Modulation
THD	Total Harmonic Distortion
UPS	Uninterruptable Power Supplies
VPE	Vacuum Pressure Encapsulation
VPI	Vacuum Pressure Impregnation
VSD	Variable Speed Drive
VSI	Voltage Source Inverters

Preface

Power Electronics is the study of switching electronic circuits in order to control the flow of electrical energy. Power Electronics is the technology behind switching power supplies, power converters, power inverters, motor drives, and motor soft starters. The first high power electronic devices were mercury-arc valves. In modern systems the conversion is performed with semiconductor switching devices such as diodes, thyristors and transistors. In contrast to electronic systems concerned with transmission and processing of signals and data, in power electronics substantial amounts of electrical energy are processed. An AC/DC converter (rectifier) is the most typical power electronics device found in many consumer electronic devices, e.g. television sets, personal computers, battery chargers, etc. The power range is typically from tens of watts to several hundred watts. In industry a common application is the variable speed drive (VSD) that is used to control an induction motor. The power range of VSDs start from a few hundred watts and end at tens of megawatts. Power electronics started with the development of the mercury arc rectifier. Invented by Peter Cooper Hewitt in 1902, it was used to convert alternating current (AC) into direct current (DC). From the 1920s on, research continued on applying thyratrons and grid-controlled mercury arc valves to power transmission. Uno Lamm developed a mercury valve with grading electrodes making them suitable for high voltage direct current power transmission. In 1933 selenium rectifiers were invented. By the 1960s the improved switching speed of bipolar junction transistors had allowed for high frequency DC/DC converters. In 1976 power MOSFETs became commercially available. In 1982 the Insulated Gate Bipolar Transistor (IGBT) was introduced. This Text aims to introduce and explain the main concepts in Power Electronics, which include Power Semi-Conductor Devices, Phase-Controlled Converters, DC to DC Converter, Inverters and AC to AC Converters. It provides a basic knowledge of circuitry for the control and conversion of electrical power with high efficiency. These converters can change and regulate the voltage, current, or power; dc-dc converters, ac-dc rectifiers, dc-ac inverters, and ac-ac cycloconverters are in common use. Applications include electronic power supplies, aerospace and vehicular power systems, and renewable energy systems.

Content Coverage

Chapter One is intended to present the fundamentals of power electronics systems. Power electronic devices may be used as switches, or as amplifiers. An ideal switch is either open or closed and so dissipates no power; it withstands an applied voltage and passes no current, or passes any amount of current with no voltage drop.

Chapter Two focuses on power semiconductor devices used as switches or rectifiers in power electronic circuits. They are also called power devices or when used in integrated circuits, called power ICs.

Chapter Three explores the use of chopper switches, a chopper circuit is used to refer to numerous types of electronic switching devices and circuits used in power control and signal applications.

Chapters Four and *Five* highlight the DC to DC and DC to AC converters, respectively. A DC-DC converter is a type of electronic circuit that takes the DC source and then converts it from the current voltage into another voltage. Devices used to convert D.C. to A.C. are called inverters.

Chapter Six focuses on AC to AC converters that convert AC waveforms to another AC waveform. The output voltage and frequency can be managed according to requirements and usage.

Chapter Seven explores the rectifier devices that converts alternating current (AC), which periodically reverses direction, to direct current (DC), which flows in only one direction.

Chapter Eight presents an overview of switched-mode power supply (SMPS) that converts power using switching devices that are turned on and off at high frequencies, and storage components such as inductors or capacitors to supply power when the switching device is in its non-conduction state.

CHAPTER 1

The Power Electronics System

Chapter Contents

- **Introduction**
- **History**
- **Devices**
- **Solid-State Devices**
- **Simulations of Power Electronic Systems**
- **Applications**

Power electronics refers to an interdisciplinary subject within electrical engineering that deals with the design, control and conversion of power in its electric form. A system that converts electric energy to an electric load through a control circuit is known as a Power Electronic System.

INTRODUCTION

Power electronics is the application of solid-state electronics to the control and conversion of electric power. It also refers to a subject of research in electronic and electrical engineering which deals with the design, control, computation and integration of nonlinear, time-varying energy-processing electronic systems with fast dynamics.

The first high power electronic devices were mercury-arc valves. In modern systems the conversion is performed with semiconductor switching devices such as diodes, thyristors and transistors, pioneered by R. D. Middlebrook and others beginning in the 1950s. In contrast to electronic systems concerned with transmission and processing of signals and data, in power electronics substantial amounts of electrical energy are processed. An AC/DC converter (rectifier) is the most typical power electronics device found in many consumer electronic devices, e.g. television sets, personal computers, battery chargers, etc. The power range is typically from tens of watts to several hundred watts. In industry a common application is the variable speed drive (VSD) that is used to control an induction motor. The power range of VSDs start from a few hundred watts and end at tens of megawatts.

The power conversion systems can be classified according to the type of the input and output power

- AC to DC (rectifier)
- DC to AC (inverter)
- DC to DC (DC-to-DC converter)
- AC to AC (AC-to-AC converter)

An HVDC thyristor valve tower 16.8 m tall in a hall at Baltic Cable AB in Sweden

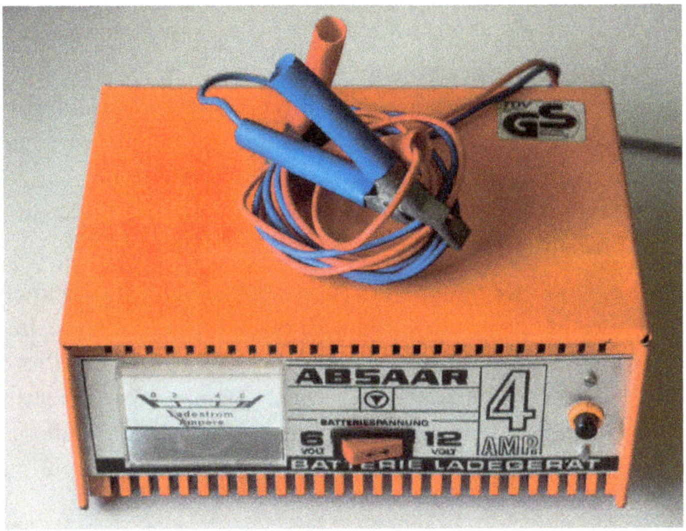

A battery charger is an example of a piece of power electronics

A PCs power supply is an example of a piece of power electronics, whether inside or outside of the cabinet.

HISTORY

Power electronics started with the development of the mercury arc rectifier. Invented by Peter Cooper Hewitt in 1902, it was used to convert alternating current (AC) into direct current (DC). From the 1920s on, research continued on applying thyratrons and grid-controlled mercury arc valves to power transmission. Uno Lamm developed a mercury valve with grading electrodes making them suitable for high voltage direct current power transmission. In 1933 selenium rectifiers were invented.

In 1947 the bipolar point-contact transistor was invented by Walter H. Brattain and John Bardeen under the direction of William Shockley at Bell Labs. In 1948 Shockley's invention of the bipolar junction transistor (BJT) improved the stability and performance of transistors, and reduced costs. By the 1950s, higher power semiconductor diodes became available and started replacing vacuum tubes. In 1956 the silicon controlled rectifier (SCR) was introduced by General Electric, greatly increasing the range of power electronics applications.

By the 1960s the improved switching speed of bipolar junction transistors had allowed for high frequency DC/DC converters. In 1976 power MOSFETs became commercially available. In 1982 the Insulated Gate Bipolar Transistor (IGBT) was introduced.

DEVICES

The capabilities and economy of power electronics system are determined by the active devices that are available. Their characteristics and limitations are a key element in the design of power electronics systems. Formerly, the mercury arc valve, the high-vacuum and gas-filled diode thermionic rectifiers, and triggered devices such as the thyratron and ignitron were widely used in power electronics. As the ratings of solid-state devices improved in both voltage and current-handling capacity, vacuum devices have been nearly entirely replaced by solid-state devices.

Power electronic devices may be used as switches, or as amplifiers. An ideal switch is either open or closed and so dissipates no power; it withstands an applied voltage and passes no current, or passes any amount of current with no voltage drop. Semiconductor devices used as switches can approximate this ideal property and so most power electronic applications rely on switching devices on and off, which makes systems very efficient as very little power is wasted in the switch. By contrast, in the case of the amplifier, the current through the device varies continuously according to a controlled input. The voltage and current at the device terminals follow a load line, and the power dissipation inside the device is large compared with the power delivered to the load.

Several attributes dictate how devices are used. Devices such as diodes conduct when a forward voltage is applied and have no external control of the start of conduction. Power devices such as silicon controlled rectifiers and thyristors (as well as the mercury valve and thyratron) allow control of the start of conduction, but rely on periodic reversal of current flow to turn them off. Devices such as gate turn-off thyristors, BJT and MOSFET transistors provide full switching control and can be turned on or off without regard to the current flow through them. Transistor devices also allow proportional amplification, but this is rarely used for systems rated more than a few hundred watts. The control input characteristics of a device also greatly affect design; sometimes the control input is at a very high voltage with respect to ground and must be driven by an isolated source.

As efficiency is at a premium in a power electronic converter, the losses that a power electronic device generates should be as low as possible.

Devices vary in switching speed. Some diodes and thyristors are suited for relatively slow speed and are useful for power frequency switching and control; certain thyristors are useful at a few kilohertz. Devices such as MOSFETS and BJTs can switch at tens of kilohertz up to a few megahertz in power applications, but with decreasing power levels. Vacuum tube devices dominate high power (hundreds of kilowatts) at very high frequency (hundreds or thousands of megahertz) applications. Faster switching devices minimize energy lost in the transitions from on to off and back, but may create problems with radiated electromagnetic interference. Gate drive (or equivalent) circuits must be designed to supply sufficient drive current to achieve the full switching speed possible with a device. A device without sufficient drive to switch rapidly may be destroyed by excess heating.

Practical devices have non-zero voltage drop and dissipate power when on, and take some time to pass through an active region until they reach the "on" or "off" state. These losses are a significant part of the total lost power in a converter.

Power handling and dissipation of devices is also a critical factor in design. Power electronic devices may have to dissipate tens or hundreds of watts of waste heat, even switching as efficiently as possible between conducting and non-conducting states. In the switching mode, the power controlled is much larger than the power dissipated in the switch. The forward voltage drop in the conducting state translates into heat that must be dissipated. High power semiconductors require specialized heat sinks or active cooling systems to manage their junction temperature; exotic semiconductors such as silicon carbide have an advantage over straight silicon in this respect, and germanium, once the main-stay of solid-state electronics is now little used due to its unfavorable high temperature properties.

Semiconductor devices exist with ratings up to a few kilovolts in a single device. Where very high voltage must be controlled, multiple devices must be used in series, with networks to equalize voltage across all devices. Again, switching speed is a critical factor since the slowest-switching device will have to withstand a disproportionate share of the overall voltage. Mercury valves were once available with ratings to 100 kV in a single unit, simplifying their application in HVDC systems.

The current rating of a semiconductor device is limited by the heat generated within the dies and the heat developed in the resistance of the interconnecting leads. Semiconductor devices must be designed so that current is evenly distributed within the device across its internal junctions (or channels); once a "hot spot" develops, breakdown effects can rapidly destroy the device. Certain SCRs are available with current ratings to 3000 amperes in a single unit.

SOLID-STATE DEVICES

Device	Description	Ratings
Diode	Uni-polar, uncontrolled, switching device used in applications such as rectification and circuit directional current control. Reverse voltage blocking device, commonly modeled as a switch in series with a voltage source, usually 0.7 VDC. The model can be enhanced to include a junction resistance, in order to accurately predict the diode voltage drop across the diode with respect to current flow.	Up to 3000 amperes and 5000 volts in a single silicon device. High voltage requires multiple series silicon devices.
Silicon-controlled rectifier (SCR)	This semi-controlled device turns on when a gate pulse is present and the anode is positive compared to the cathode. When a gate pulse is present, the device operates like a standard diode. When the anode is negative compared to the cathode, the device turns off and blocks positive or negative voltages present. The gate voltage does not allow the device to turn off.	Up to 3000 amperes, 5000 volts in a single silicon device.
Thyristor	The thyristor is a family of three-terminal devices that include SCRs, GTOs, and MCT. For most of the devices, a gate pulse turns the device on. The device turns off when the anode voltage falls below a value (relative to the cathode) determined by the device characteristics. When off, it is considered a reverse voltage blocking device.	
Gate turn-off thyristor (GTO)	The gate turn-off thyristor, unlike an SCR, can be turned on and off with a gate pulse. One issue with the device is that turn off gate voltages are usually larger and require more current than turn on levels. This turn off voltage is a negative voltage from gate to source, usually it only needs to be present for a short time, but the magnitude s on the order of 1/3 of the anode current. A snubber circuit is required in order to provide a usable switching curve for this device. Without the snubber circuit, the GTO cannot be used for turning inductive loads off. These devices, because of developments in IGCT technology are not very popular in the power electronics realm. They are considered controlled, uni-polar and bi-polar voltage blocking.	
Triac	The triac is a device that is essentially an integrated pair of phase-controlled thyristors connected in inverse-parallel on the same chip. Like an SCR, when a voltage pulse is present on the gate terminal, the device turns on. The main difference between an SCR and a Triac is that both the positive and negative cycle can be turned on independently of each other, using a positive or negative gate pulse. Similar to an SCR, once the device is turned on, the device cannot be turned off. This device is considered bi-polar and reverse voltage blocking.	

Bipolar junction transistor (BJT)	The BJT cannot be used at high power; they are slower and have more resistive losses when compared to MOSFET type devices. To carry high current, BJTs must have relatively large base currents, thus these devices have high power losses when compared to MOSFET devices. BJTs along with MOSFETs, are also considered unipolar and do not block reverse voltage very well, unless installed in pairs with protection diodes. Generally, BJTs are not utilized in power electronics switching circuits because of the I^2R losses associated with on resistance and base current requirements. BJTs have lower current gains in high power packages, thus requiring them to be set up in Darlington configurations in order to handle the currents required by power electronic circuits. Because of these multiple transistor configurations, switching times are in the hundreds of nanoseconds to microseconds. Devices have voltage ratings which max out around 1500 V and fairly high current ratings. They can also be paralleled in order to increase power handling, but must be limited to around 5 devices for current sharing.	
Power MOSFET	The main benefit of the power MOSFET is that the base current for BJT is large compared to almost zero for MOSFET gate current. Since the MOSFET is a depletion channel device, voltage, not current, is necessary to create a conduction path from drain to source. The gate does not contribute to either drain or source current. Turn on gate current is essentially zero with the only power dissipated at the gate coming during switching. Losses in MOSFETs are largely attributed to on-resistance. The calculations show a direct correlation to drain source on-resistance and the device blocking voltage rating, BV_{dss}. Switching times range from tens of nanoseconds to a few hundred microseconds, depending on the device. MOSFET drain source resistances increase as more current flows through the device. As frequencies increase the losses increase as well, making BJTs more attractive. Power MOSFETs can be paralleled in order to increase switching current and therefore overall switching power. Nominal voltages for MOSFET switching devices range from a few volts to a little over 1000 V, with currents up to about 100 A or so. Newer devices may have higher operational characteristics. MOSFET devices are not bi-directional, nor are they reverse voltage blocking. ||	
Insulated-gate bipolar transistor (IGBT)	These devices have the best characteristics of MOSFETs and BJTs. Like MOSFET devices, the insulated gate bipolar transistor has a high gate impedance, thus low gate current requirements. Like BJTs, this device has low on state voltage drop, thus low power loss across the switch in operating mode. Similar to the GTO, the IGBT can be used to block both positive and negative voltages. Operating currents are fairly high, in excess of 1500 A and switching voltage up to 3000 V. The IGBT has reduced input capacitance compared to MOSFET devices which improves the Miller feedback effect during high dv/dt turn on and turn off.	

MOS-controlled thyristor (MCT)	The MOS-controlled thyristor is thyristor like and can be triggered on or off by a pulse to the MOSFET gate. Since the input is MOS technology, there is very little current flow, allowing for very low power control signals. The device is constructed with two MOSFET inputs and a pair of BJT output stages. Input MOSFETs are configured to allow turn on control during positive and negative half cycles. The output BJTs are configured to allow for bidirectional control and low voltage reverse blocking. Some benefits to the MCT are fast switching frequencies, fairly high voltage and medium current ratings (around 100 A or so).	
Integrated gate-commutated thyristor (IGCT)	Similar to a GTO, but without the high current requirements to turn on or off the load. The IGCT can be used for quick switching with little gate current. The devices high input impedance largely because of the MOSFET gate drivers. They have low resistance outputs that don't waste power and very fast transient times that rival that of BJTs. ABB Group company has published data sheets for these devices and provided descriptions of the inner workings. The device consists of a gate, with an optically isolated input, low on resistance BJT output transistors which lead to a low voltage drop and low power loss across the device at fairly high switching voltage and current levels. An example of this new device from ABB shows how this device improves on GTO technology for switching high voltage and high current in power electronics applications. According to ABB, the IGCT devices are capable of switching in excess of 5000 VAC and 5000 A at very high frequencies, something not possible to do efficiently with GTO devices.	

DC/AC Converters (Inverters)

DC to AC converters produce an AC output waveform from a DC source. Applications include adjustable speed drives (ASD), uninterruptable power supplies (UPS), active filters, Flexible AC transmission systems (FACTS), voltage compensators, and photovoltaic generators. Topologies for these converters can be separated into two distinct categories: voltage source inverters and current source inverters. Voltage source inverters (VSIs) are named so because the independently controlled output is a voltage waveform. Similarly, current source inverters (CSIs) are distinct in that the controlled AC output is a current waveform.

Being static power converters, the DC to AC power conversion is the result of power switching devices, which are commonly fully controllable semiconductor power switches. The output waveforms are therefore made up of discrete values, producing fast transitions rather than smooth ones. The ability to produce near sinusoidal waveforms around the fundamental frequency is dictated by the modulation technique controlling when, and for how long, the power valves are on and off. Common modulation techniques include the carrier-based technique, or Pulse-width modulation, space-vector technique, and the selective-harmonic technique.

Voltage source inverters have practical uses in both single-phase and three-phase applications. Single-phase VSIs utilize half-bridge and full-bridge configurations, and are widely used for power supplies, single-phase UPSs, and elaborate high-power topologies when used in multicell configurations. Three-phase VSIs are used in applications that require sinusoidal voltage waveforms, such as ASDs, UPSs, and some types of FACTS devices such as the STATCOM. They are also used in applications where arbitrary voltages are required as in the case of active filters and voltage compensators.

Current source inverters are used to produce an AC output current from a DC current supply. This type of inverter is practical for three-phase applications in which high-quality voltage waveforms are required.

A relatively new class of inverters, called multilevel inverters, has gained widespread interest. Normal operation of CSIs and VSIs can be classified as two-level inverters, due to the fact that power switches connect to either the positive or to the negative DC bus. If more than two voltage levels were available to the inverter output terminals, the AC output could better approximate a sine wave. It is for this reason that multilevel inverters, although more complex and costly, offer higher performance.

Each inverter type differs in the DC links used, and in whether or not they require freewheeling diodes. Either can be made to operate in square-wave or pulse-width modulation (PWM) mode, depending on its intended usage. Square-wave mode offers simplicity, while PWM can be implemented several different ways and produces higher quality waveforms.

Voltage Source Inverters (VSI) feed the output inverter section from an approximately constant-voltage source.

The desired quality of the current output waveform determines which modulation technique needs to be selected for a given application. The output of a VSI is composed of discrete values. In order to obtain a smooth current waveform, the loads need to be inductive at the select harmonic frequencies. Without some sort of inductive filtering between the source and load, a capacitive load will cause the load to receive a choppy current waveform, with large and frequent current spikes.

There are three main types of VSIs:

- Single-phase half-bridge inverter
- Single-phase full-bridge inverter
- Three-phase voltage source inverter

Single-Phase Half-Bridge Inverter

The single-phase voltage source half-bridge inverters, are meant for lower voltage applications and are commonly used in power supplies.

Low-order current harmonics get injected back to the source voltage by the operation of the inverter. This means that two large capacitors are needed for filtering purposes in this design. As figure below illustrates, only one switch can be on at time in each leg of the inverter. If both switches in a leg were on at the same time, the DC source will be shorted out.

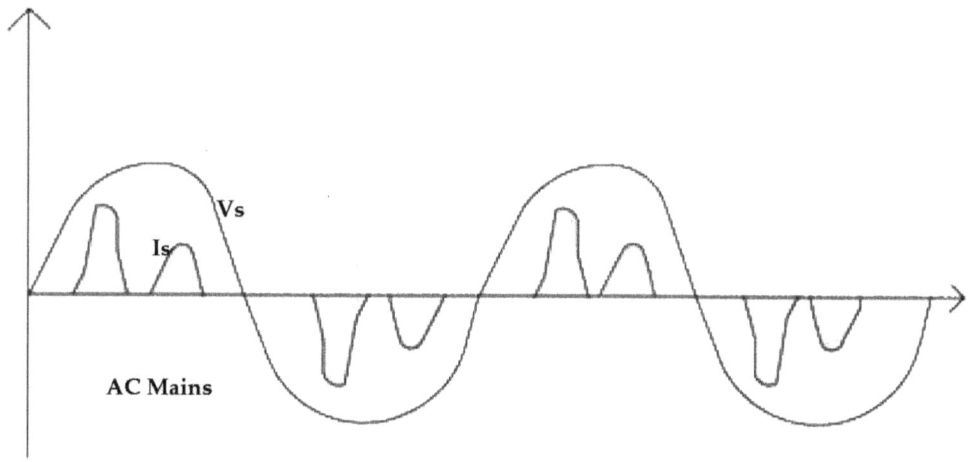

The AC input for an ASD.

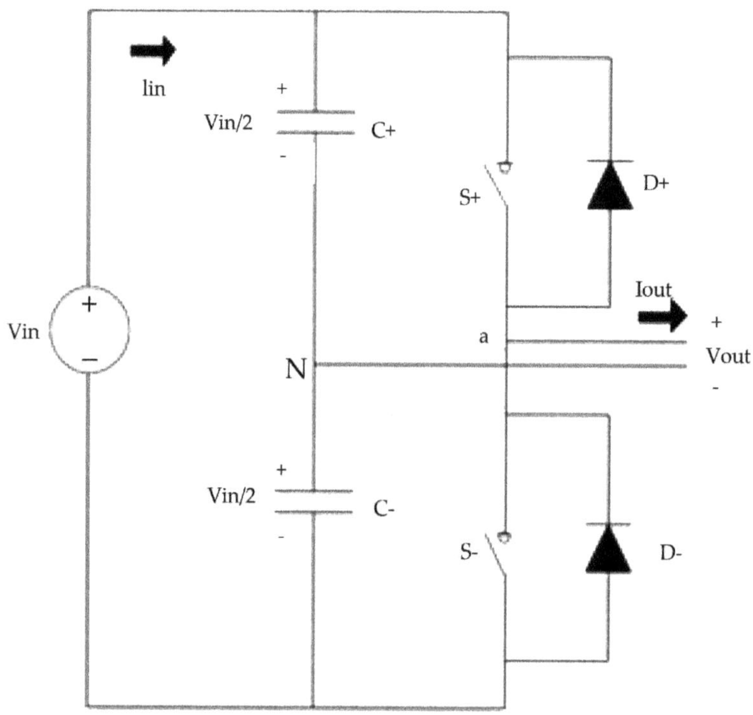

Single-Phase Half-Bridge Voltage Source Inverter.

Inverters can use several modulation techniques to control their switching schemes. The carrier-based PWM technique compares the AC output waveform, v_c, to a carrier voltage signal, v_Δ. When v_c is greater than v_Δ, S+ is on, and when v_c is less than v_Δ, S- is on. When the AC output is at frequency fc with its amplitude at v_c, and the triangular carrier signal is at frequency f_Δ with its amplitude at v_Δ, the PWM becomes a special sinusoidal case of the carrier based PWM. This case is dubbed sinusoidal pulse-width modulation (SPWM).For this, the modulation index, or amplitude-modulation ratio, is defined as $\mathbf{m_a = v_c/v_\Delta}$.

The normalized carrier frequency, or frequency-modulation ratio, is calculated using the equation $\mathbf{m_f = f_\Delta/f_c}$.

If the over-modulation region, ma, exceeds one, a higher fundamental AC output voltage will be observed, but at the cost of saturation. For SPWM, the harmonics of the output waveform are at well-defined frequencies and amplitudes. This simplifies the design of the filtering components needed for the low-order current harmonic injection from the operation of the inverter. The maximum output amplitude in this mode of operation is half of the source voltage. If the maximum output amplitude, m_a, exceeds 3.24, the output waveform of the inverter becomes a square wave.

As was true for PWM, both switches in a leg for square wave modulation cannot be turned on at the same time, as this would cause a short across the voltage source. The switching scheme requires that both S+ and S- be on for a half cycle of the AC output period. The fundamental AC output amplitude is equal to $\mathbf{v_{o1} = v_{aN} = 2v_i/\pi}$.

Its harmonics have an amplitude of $\mathbf{v_{oh} = v_{o1}/h}$.

Therefore, the AC output voltage is not controlled by the inverter, but rather by the magnitude of the DC input voltage of the inverter.

Using selective harmonic elimination (SHE) as a modulation technique allows the switching of the inverter to selectively eliminate intrinsic harmonics. The fundamental component of the AC output voltage can also be adjusted within a desirable range. Since the AC output voltage obtained from this modulation technique has odd half and odd quarter wave symmetry, even harmonics do not exist. Any undesirable odd (N-1) intrinsic harmonics from the output waveform can be eliminated.

Single-Phase Full-Bridge Inverter

The full-bridge inverter is similar to the half bridge-inverter, but it has an additional leg to connect the neutral point to the load. Figure below shows the circuit schematic of the single-phase voltage source full-bridge inverter.

To avoid shorting out the voltage source, S1+ and S1- cannot be on at the same time, and S2+ and S2- also cannot be on at the same time. Any modulating technique used for the full-bridge configuration should have either the top or the bottom switch of each leg on at any given time. Due to the extra leg, the maximum amplitude of the output waveform is Vi, and is twice as large as the maximum achievable output amplitude for the half-bridge configuration.

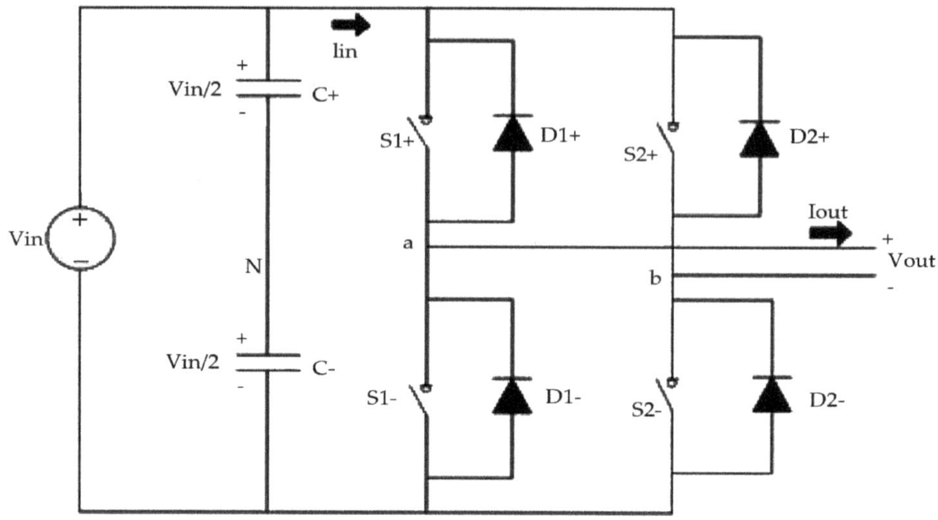

Single-Phase Voltage Source Full-Bridge Inverter

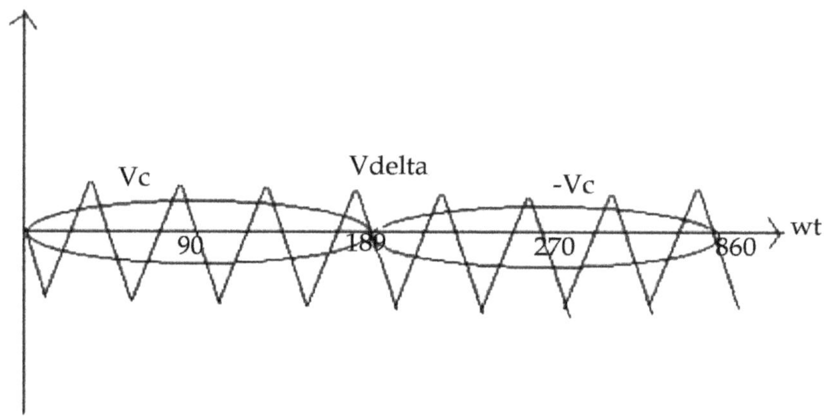

Carrier and Modulating Signals for the Bipolar Pulsewidth Modulation Technique.

States 1 and 2 from Table 2 are used to generate the AC output voltage with bipolar SPWM. The AC output voltage can take on only two values, either Vi or –Vi. To generate these same states using a half-bridge configuration, a carrier based technique can be used. S+ being on for the half-bridge corresponds to S1+ and S2- being on for the full-bridge. Similarly, S- being on for the half-bridge corresponds to S1- and S2+ being on for the full bridge. The output voltage for this modulation technique is more or less sinusoidal, with a fundamental component that has an amplitude in the linear region of less than or equal to one $\mathbf{v}_{o1} = \mathbf{v}_{ab1} = \mathbf{v}_i \bullet \mathbf{m}_a$.

Unlike the bipolar PWM technique, the unipolar approach uses states 1, 2, 3 and 4 from Table 2 to generate its AC output voltage. Therefore, the AC output voltage can take on the values Vi, 0 or –V i. To generate these states, two sinusoidal modulating signals, Vc and –Vc, are needed.

Vc is used to generate VaN, while –Vc is used to generate VbN. The following relationship is called unipolar carrier-based SPWM $\mathbf{v_{o1} = 2 \cdot v_{aN1} = v_i \cdot m_a}$.

The phase voltages VaN and VbN are identical, but 180 degrees out of phase with each other. The output voltage is equal to the difference of the two phase voltages, and do not contain any even harmonics. Therefore, if mf is taken, even the AC output voltage harmonics will appear at normalized odd frequencies, fh. These frequencies are centered on double the value of the normalized carrier frequency. This particular feature allows for smaller filtering components when trying to obtain a higher quality output waveform.

As was the case for the half-bridge SHE, the AC output voltage contains no even harmonics due to its odd half and odd quarter wave symmetry.

Three-Phase Voltage Source Inverter

Single-phase VSIs are used primarily for low power range applications, while three-phase VSIs cover both medium and high power range applications.

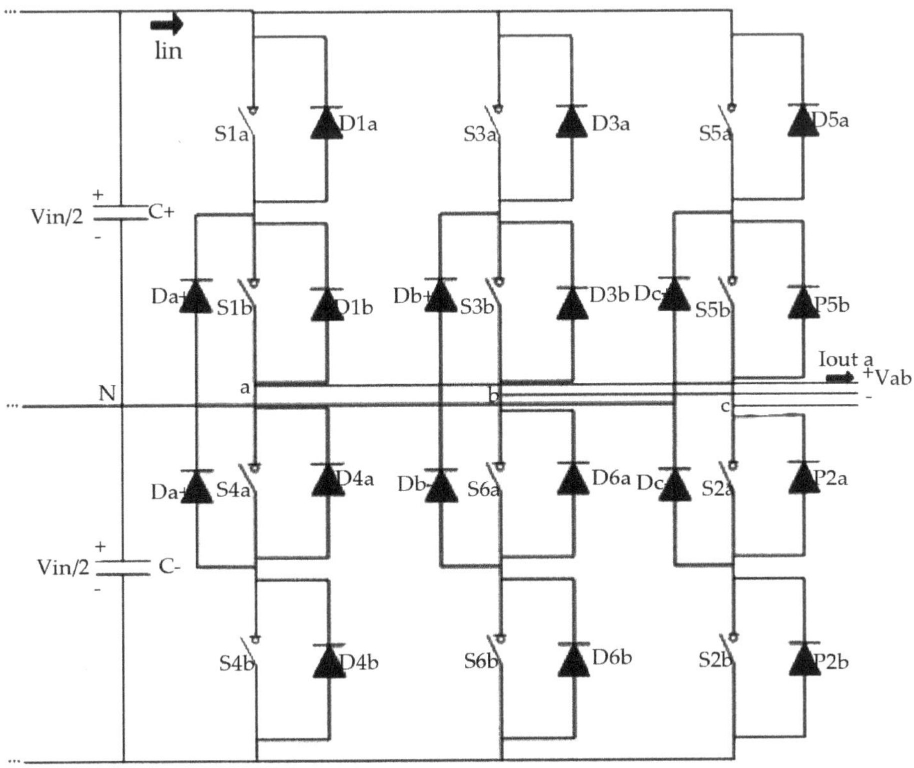

Three-Phase Voltage Source Inverter Circuit Schematic

Switches in any of the three legs of the inverter cannot be switched off simultaneously due to this resulting in the voltages being dependent on the respective line current's polarity. States 7 and 8 produce zero AC line voltages, which result in AC line currents freewheeling through either the upper or the lower components. However, the line voltages for states 1 through 6 produce an AC line voltage consisting of the discrete values of Vi, 0 or –Vi. For three-phase SPWM, three modulating signals that are 120 degrees out of phase with one another are used in order to produce out of phase load voltages. In order to preserve the PWM features with a single carrier signal, the normalized carrier frequency, mf, needs to be a multiple of three. This keeps the magnitude of the phase voltages identical, but out of phase with each other by 120 degrees. The maximum achievable phase voltage amplitude in the linear region, ma less than or equal to one, is $\mathbf{v_{phase} = v_i / 2}$. The maximum achievable line voltage amplitude is $\mathbf{V_{ab1} = v_{ab} \cdot \sqrt{3} / 2}$

The only way to control the load voltage is by changing the input DC voltage.

Current Source Inverters

Current source inverters convert DC current into an AC current waveform. In applications requiring sinusoidal AC waveforms, magnitude, frequency, and phase should all be controlled. CSIs have high changes in current overtime, so capacitors are commonly employed on the AC side, while inductors are commonly employed on the DC side. Due to the absence of freewheeling diodes, the power circuit is reduced in size and weight, and tends to be more reliable than VSIs. Although single-phase topologies are possible, three-phase CSIs are more practical.

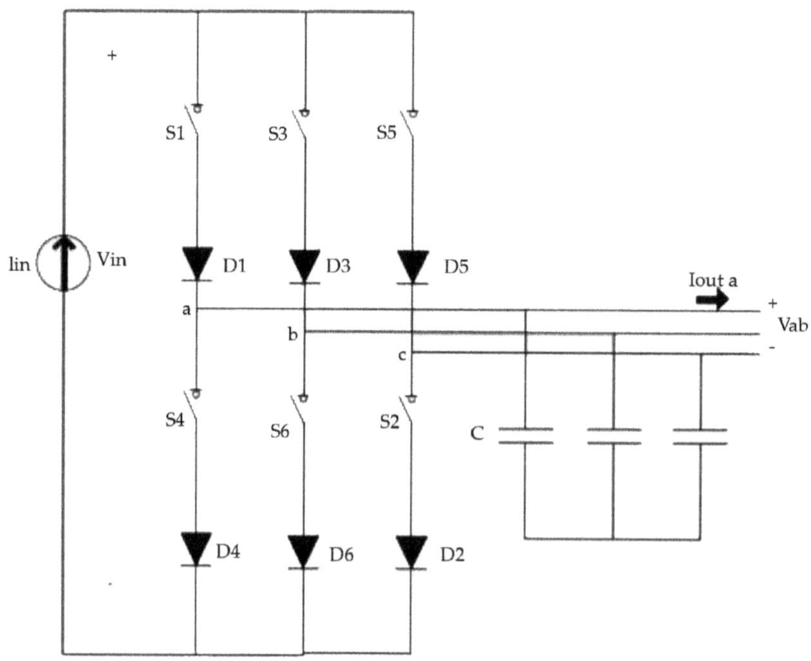

Three-Phase Current Source Inverter.

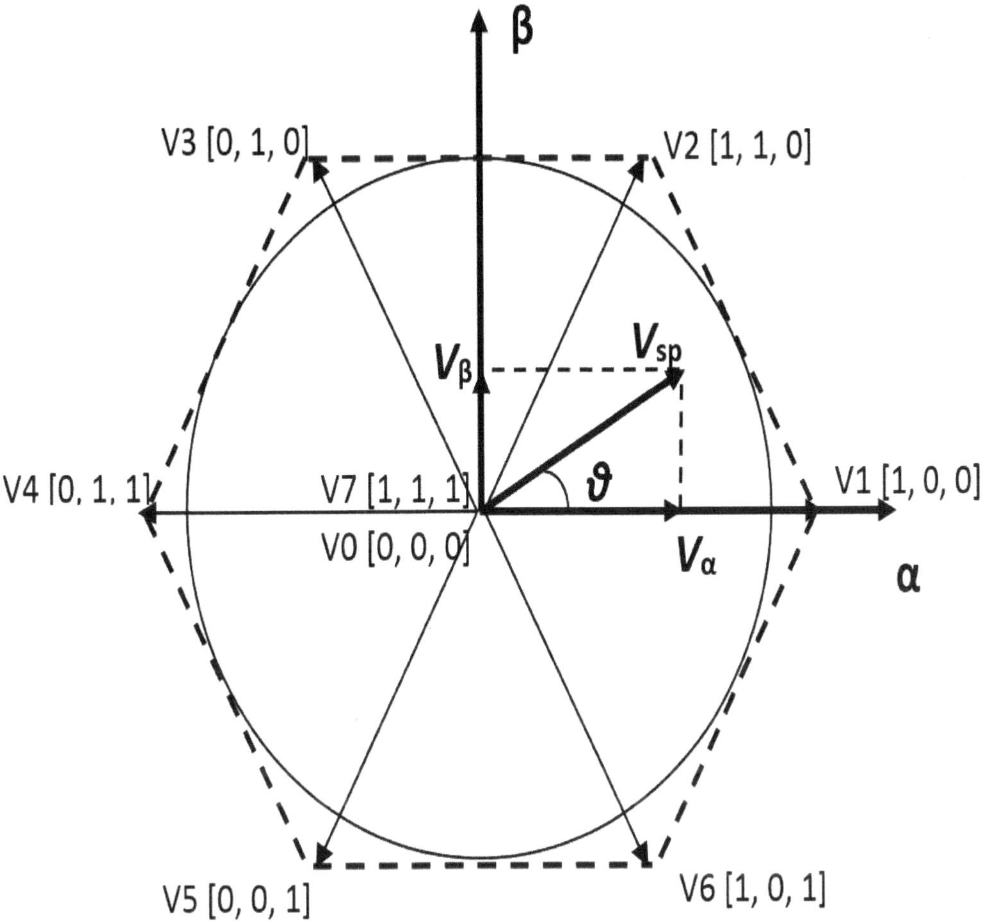

Space-Vector Representation in Current Source Inverters.

In its most generalized form, a three-phase CSI employs the same conduction sequence as a six-pulse rectifier. At any time, only one common-cathode switch and one common-anode switch are on.

As a result, line currents take discrete values of –ii, 0 and ii. States are chosen such that a desired waveform is outputted and only valid states are used. This selection is based on modulating techniques, which include carrier-based PWM, selective harmonic elimination, and space-vector techniques.

Carrier-based techniques used for VSIs can also be implemented for CSIs, resulting in CSI line currents that behave in the same way as VSI line voltages. The digital circuit utilized for modulating signals contains a switching pulse generator, a shorting pulse generator, a shorting pulse distributor, and a switching and shorting pulse combiner. A gating signal is produced based on a carrier current and three modulating signals.

A shorting pulse is added to this signal when no top switches and no bottom switches are gated, causing the RMS currents to be equal in all legs. The same methods are utilized for each phase, however, switching variables are 120 degrees out of phase relative to one another, and the current pulses are shifted by a half-cycle with respect to output currents. If a triangular carrier is used with sinusoidal modulating signals, the CSI is said to be utilizing synchronized-pulse-width-modulation (SPWM). If full over-modulation is used in conjunction with SPWM the inverter is said to be in square-wave operation.

The second CSI modulation category, SHE is also similar to its VSI counterpart. Utilizing the gating signals developed for a VSI and a set of synchronizing sinusoidal current signals, results in symmetrically distributed shorting pulses and, therefore, symmetrical gating patterns. This allows any arbitrary number of harmonics to be eliminated. It also allows control of the fundamental line current through the proper selection of primary switching angles. Optimal switching patterns must have quarter-wave and half-wave symmetry, as well as symmetry about 30 degrees and 150 degrees. Switching patterns are never allowed between 60 degrees and 120 degrees. The current ripple can be further reduced with the use of larger output capacitors, or by increasing the number of switching pulses.

The third category, space-vector-based modulation, generates PWM load line currents that equal load line currents, on average. Valid switching states and time selections are made digitally based on space vector transformation. Modulating signals are represented as a complex vector using a transformation equation. For balanced three-phase sinusoidal signals, this vector becomes a fixed module, which rotates at a frequency, ω. These space vectors are then used to approximate the modulating signal. If the signal is between arbitrary vectors, the vectors are combined with the zero vectors I7, I8, or I9. The following equations are used to ensure that the generated currents and the current vectors are on average equivalent.

Multilevel Inverters

A relatively new class called multilevel inverters has gained widespread interest. Normal operation of CSIs and VSIs can be classified as two-level inverters because the power switches connect to either the positive or the negative DC bus. If more than two voltage levels were available to the inverter output terminals, the AC output could better approximate a sine wave. For this reason multilevel inverters, although more complex and costly, offer higher performance.

Control methods for a three-level inverter only allow two switches of the four switches in each leg to simultaneously change conduction states. This allows smooth commutation and avoids shoot through by only selecting valid states. It may also be noted that since the DC bus voltage is shared by at least two power valves, their voltage ratings can be less than a two-level counterpart.

Carrier-based and space-vector modulation techniques are used for multilevel topologies. The methods for these techniques follow those of classic inverters, but with added complexity. Space-vector modulation offers a greater number of fixed voltage vectors to be used in approximating the modulation signal, and therefore allows more effective space vector PWM strategies to be accomplished at the cost of more elaborate algorithms. Due to added complexity and number of semiconductor devices, multilevel inverters are currently more suitable for high-power high-voltage applications. This technology reduces the harmonics hence improves overall efficiency of the scheme.

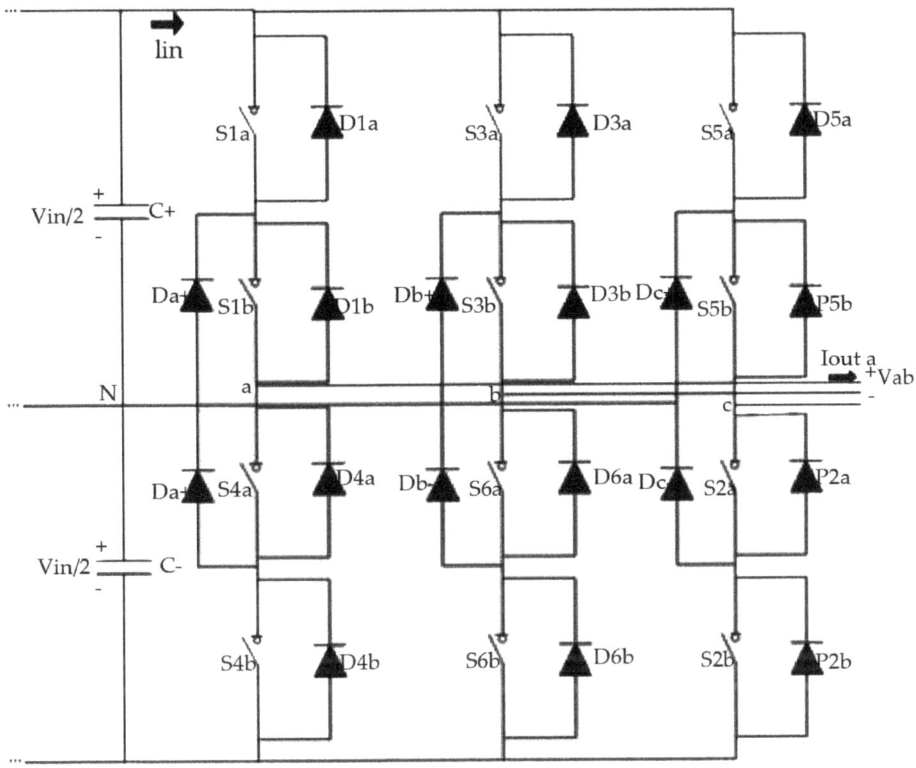

Three-Level Neutral-Clamped Inverter

AC/AC Converters

Converting AC power to AC power allows control of the voltage, frequency, and phase of the waveform applied to a load from a supplied AC system. The two main categories that can be used to separate the types of converters are whether the frequency of the waveform is changed. AC/AC converter that don't allow the user to modify the frequencies are known as AC Voltage Controllers, or AC Regulators. AC converters that allow the user to change the frequency are simply referred to as frequency converters for AC to AC conversion. Under frequency converters there are three different types of converters that are typically used: cycloconverter, matrix converter, DC link converter (aka AC/DC/AC converter).

AC voltage controller: The purpose of an AC Voltage Controller, or AC Regulator, is to vary the RMS voltage across the load while at a constant frequency. Three control methods that are generally accepted are ON/OFF Control, Phase-Angle Control, and Pulse Width Modulation AC Chopper Control (PWM AC Chopper Control). All three of these methods can be implemented not only in single-phase circuits, but three-phase circuits as well.

- ON/OFF Control: Typically used for heating loads or speed control of motors, this control method involves turning the switch on for n integral cycles and turning the switch off for m integral cycles. Because turning the switches on and off causes undesirable harmonics to be created, the switches are turned on and off during zero-voltage and zero-current conditions (zero-crossing), effectively reducing the distortion.
- Phase-Angle Control: Various circuits exist to implement a phase-angle control on different waveforms, such as half-wave or full-wave voltage control. The power electronic components that are typically used are diodes, SCRs, and Triacs. With the use of these components, the user can delay the firing angle in a wave which will only cause part of the wave to be outputted.
- PWM AC Chopper Control: The other two control methods often have poor harmonics, output current quality, and input power factor. In order to improve these values PWM can be used instead of the other methods. What PWM AC Chopper does is have switches that turn on and off several times within alternate half-cycles of input voltage.

Matrix converters and cycloconverters: Cycloconverters are widely used in industry for ac to ac conversion, because they are able to be used in high-power applications. They are commutated direct frequency converters that are synchronised by a supply line. The cycloconverters output voltage waveforms have complex harmonics with the higher order harmonics being filtered by the machine inductance. Causing the machine current to have fewer harmonics, while the remaining harmonics causes losses and torque pulsations. Note that in a cycloconverter, unlike other converters, there are no inductors or capacitors, i.e. no storage devices. For this reason, the instantaneous input power and the output power are equal.

- Single-Phase to Single-Phase Cycloconverters: Single-Phase to Single-Phase Cycloconverters started drawing more interest recently because of the decrease in both size and price of the power electronics switches. The single-phase high frequency ac voltage can be either sinusoidal or trapezoidal. These might be zero voltage intervals for control purpose or zero voltage commutation.
- Three-Phase to Single-Phase Cycloconverters: There are two kinds of three-phase to single-phase cycloconverters: 3φ to 1φ half wave cycloconverters and 3φ to 1φ bridge cycloconverters. Both positive and negative converters can generate voltage at either polarity, resulting in the positive converter only supplying positive current, and the negative converter only supplying negative current.

With recent device advances, newer forms of cycloconverters are being developed, such as matrix converters. The first change that is first noticed is that matrix converters utilize bi-directional, bipolar switches. A single phase to a single phase matrix converter consists of a matrix of 9 switches connecting the three input phases to the tree output phase. Any input phase and output phase can be connected together at any time without connecting any two switches from the same phase at the same time; otherwise this will cause a short circuit of the input phases. Matrix converters are lighter, more compact and versatile than other converter solutions. As a result, they are able to achieve higher levels of integration, higher temperature operation, broad output frequency and natural bi-directional power flow suitable to regenerate energy back to the utility.

The matrix converters are subdivided into two types: direct and indirect converters. A direct matrix converter with three-phase input and three-phase output, the switches in a matrix converter must be bi-directional, that is, they must be able to block voltages of either polarity and to conduct current in either direction. This switching strategy permits the highest possible output voltage and reduces the reactive line-side current. Therefore, the power flow through the converter is reversible. Because of its commutation problem and complex control keep it from being broadly utilized in industry.

Unlike the direct matrix converters, the indirect matrix converters has the same functionality, but uses separate input and output sections that are connected through a dc link without storage elements. The design includes a four-quadrant current source rectifier and a voltage source inverter. The input section consists of bi-directional bipolar switches. The commutation strategy can be applied by changing the switching state of the input section while the output section is in a freewheeling mode. This commutation algorithm is significantly less complexity and higher reliability as compared to a conventional direct matrix converter.

DC link converters: DC Link Converters, also referred to as AC/DC/AC converters, convert an AC input to an AC output with the use of a DC link in the middle. Meaning that the power in the converter is converted to DC from AC with the use of a rectifier, and then it is converted back to AC from DC with the use of an inverter. The end result is an output with a lower voltage and variable (higher or lower) frequency. Due to their wide area of application, the AC/DC/AC converters are the most common contemporary solution. Other advantages to AC/DC/AC converters is that they are stable in overload and no-load conditions, as well as they can be disengaged from a load without damage.

Hybrid matrix converter: Hybrid matrix converters are relatively new for AC/AC converters. These converters combine the AC/DC/AC design with the matrix converter design. Multiple types of hybrid converters have been developed in this new category, an example being a converter that uses uni-directional switches and two converter stages without the dc-link; without the capacitors or inductors needed for a dc-link, the weight and size of the converter is reduced. Two sub-categories exist from the hybrid converters, named hybrid direct matrix converter (HDMC) and hybrid indirect matrix converter (HIMC). HDMC convert the voltage and current in one stage, while the HIMC utilizes separate stages, like the AC/DC/AC converter, but without the use of an intermediate storage element.

Applications: Below is a list of common applications that each converter is used in.

- AC Voltage Controller: Lighting Control; Domestic and Industrial Heating; Speed Control of Fan,Pump or Hoist Drives, Soft Starting of Induction Motors, Static AC Switches(Temperature Control, Transformer Tap Changing, etc.)
- Cycloconverter: High-Power Low-Speed Reversible AC Motor Drives; Constant Frequency Power Supply with Variable Input Frequency; Controllable VAR Generators for Power Factor Correction; AC System Interties Linking Two Independent Power Systems.

- Matrix Converter: Currently the application of matrix converters are limited due to non-availability of bilateral monolithic switches capable of operating at high frequency, complex control law implementation, commutation and other reasons. With these developments, matrix converters could replace cycloconverters in many areas.
- DC Link: Can be used for individual or multiple load applications of machine building and construction.

SIMULATIONS OF POWER ELECTRONIC SYSTEMS

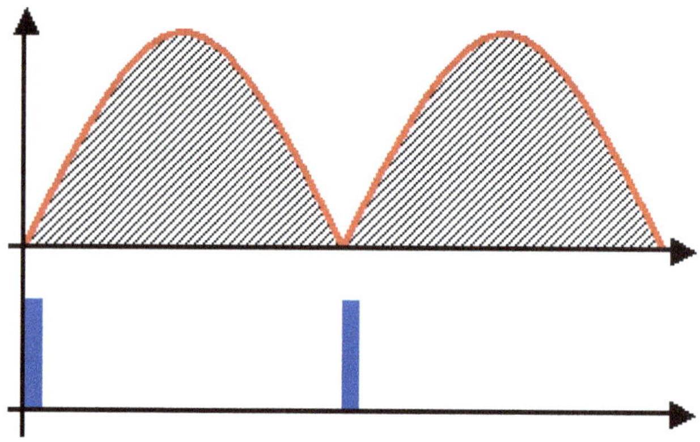

Output voltage of a full-wave rectifier with controlled thyristors

Power electronic circuits are simulated using computer simulation programs such as PLECS, PSIM and MATLAB/simulink. Circuits are simulated before they are produced to test how the circuits respond under certain conditions. Also, creating a simulation is both cheaper and faster than creating a prototype to use for testing.

APPLICATIONS

Applications of power electronics range in size from a switched mode power supply in an AC adapter, battery chargers, audio amplifiers, fluorescent lamp ballasts, through variable frequency drives and DC motor drives used to operate pumps, fans, and manufacturing machinery, up to gigawatt-scale high voltage direct current power transmission systems used to interconnect electrical grids. Power electronic systems are found in virtually every electronic device. For example:

- DC/DC converters are used in most mobile devices (mobile phones, PDA etc.) to maintain the voltage at a fixed value whatever the voltage level of the battery is. These converters are also used for electronic isolation and power factor correction. A power optimizer is a type of DC/DC converter developed to maximize the energy harvest from solar photovoltaic or wind turbine systems.

- AC/DC converters (rectifiers) are used every time an electronic device is connected to the mains (computer, television etc.). These may simply change AC to DC or can also change the voltage level as part of their operation.
- AC/AC converters are used to change either the voltage level or the frequency (international power adapters, light dimmer). In power distribution networks AC/AC converters may be used to exchange power between utility frequency 50 Hz and 60 Hz power grids.
- DC/AC converters (inverters) are used primarily in UPS or renewable energy systems or emergency lighting systems. Mains power charges the DC battery. If the mains fails, an inverter produces AC electricity at mains voltage from the DC battery. Solar inverter, both smaller string and larger central inverters, as well as solar micro-inverter are used in photovoltaics as a component of a PV system.

Motor drives are found in pumps, blowers, and mill drives for textile, paper, cement and other such facilities. Drives may be used for power conversion and for motion control. For AC motors, applications include variable-frequency drives, motor soft starters and excitation systems.

In hybrid electric vehicles (HEVs), power electronics are used in two formats: series hybrid and parallel hybrid. The difference between a series hybrid and a parallel hybrid is the relationship of the electric motor to the internal combustion engine (ICE). Devices used in electric vehicles consist mostly of dc/dc converters for battery charging and dc/ac converters to power the propulsion motor. Electric trains use power electronic devices to obtain power, as well as for vector control using pulse width modulation (PWM) rectifiers. The trains obtain their power from power lines. Another new usage for power electronics is in elevator systems. These systems may use thyristors, inverters, permanent magnet motors, or various hybrid systems that incorporate PWM systems and standard motors.

Inverters

In general, inverters are utilized in applications requiring direct conversion of electrical energy from DC to AC or indirect conversion from AC to AC. DC to AC conversion is useful for many fields, including power conditioning, harmonic compensation, motor drives, and renewable energy grid-integration.

In power systems it is often desired to eliminate harmonic content found in line currents. VSIs can be used as active power filters to provide this compensation. Based on measured line currents and voltages, a control system determines reference current signals for each phase. This is fed back through an outer loop and subtracted from actual current signals to create current signals for an inner loop to the inverter. These signals then cause the inverter to generate output currents that compensate for the harmonic content. This configuration requires no real power consumption, as it is fully fed by the line; the DC link is simply a capacitor that is kept at a constant voltage by the control system. In this configuration, output currents are in phase with line voltages to produce a unity power factor. Conversely, VAR compensation is possible in a similar configuration where output currents lead line voltages to improve the overall power factor.

In facilities that require energy at all times, such as hospitals and airports, UPS systems are utilized. In a standby system, an inverter is brought online when the normally supplying grid is interrupted. Power is instantaneously drawn from onsite batteries and converted into usable AC voltage by the VSI, until grid power is restored, or until backup generators are brought online. In an online UPS system, a rectifier-DC-link-inverter is used to protect the load from transients and harmonic content. A battery in parallel with the DC-link is kept fully charged by the output in case the grid power is interrupted, while the output of the inverter is fed through a low pass filter to the load. High power quality and independence from disturbances is achieved.

Various AC motor drives have been developed for speed, torque, and position control of AC motors. These drives can be categorized as low-performance or as high-performance, based on whether they are scalar-controlled or vector-controlled, respectively. In scalar-controlled drives, fundamental stator current, or voltage frequency and amplitude, are the only controllable quantities. Therefore, these drives are employed in applications where high quality control is not required, such as fans and compressors. On the other hand, vector-controlled drives allow for instantaneous current and voltage values to be controlled continuously. This high performance is necessary for applications such as elevators and electric cars.

Inverters are also vital to many renewable energy applications. In photovoltaic purposes, the inverter, which is usually a PWM VSI, gets fed by the DC electrical energy output of a photovoltaic module or array. The inverter then converts this into an AC voltage to be interfaced with either a load or the utility grid. Inverters may also be employed in other renewable systems, such as wind turbines. In these applications, the turbine speed usually varies causing changes in voltage frequency and sometimes in the magnitude. In this case, the generated voltage can be rectified and then inverted to stabilize frequency and magnitude.

Smart Grid

A smart grid is a modernized electrical grid that uses information and communications technology to gather and act on information, such as information about the behaviors of suppliers and consumers, in an automated fashion to improve the efficiency, reliability, economics, and sustainability of the production and distribution of electricity.

Electric power generated by wind turbines and hydroelectric turbines by using induction generators can cause variances in the frequency at which power is generated. Power electronic devices are utilized in these systems to convert the generated ac voltages into high-voltage direct current (HVDC). The HVDC power can be more easily converted into three phase power that is coherent with the power associated to the existing power grid. Through these devices, the power delivered by these systems is cleaner and has a higher associated power factor. Wind power systems optimum torque is obtained either through a gearbox or direct drive technologies that can reduce the size of the power electronics device.

Electric power can be generated through photovoltaic cells by using power electronic devices. The produced power is usually then transformed by solar inverters. Inverters are divided into three different types: central, module-integrated and string. Central converters can be connected either in parallel or in series on the DC side of the system. For photovoltaic "farms", a single central converter is used for the entire system. Module-integrated converters are connected in series on either the DC or AC side. Normally several modules are used within a photovoltaic system, since the system requires these converters on both DC and AC terminals. A string converter is used in a system that utilizes photovoltaic cells that are facing different directions. It is used to convert the power generated to each string, or line, in which the photovoltaic cells are interacting.

Grid Voltage Regulation

Power electronics can be used to help utilities adapt to the rapid increase in distributed residential/commercial solar power generation. Germany and parts of Hawaii, California and New Jersey require costly studies to be conducted before approving new solar installations. Relatively small-scale ground- or pole-mounted devices create the potential for a distributed control infrastructure to monitor and manage the flow of power. Traditional electromechanical systems, such as capacitor banks or voltage regulators at substations, can take minutes to adjust voltage and can be distant from the solar installations where the problems originate. If voltage on a neighborhood circuit goes too high, it can endanger utility crews and cause damage to both utility and customer equipment. Further, a grid fault causes photovoltaic generators to shut down immediately, spiking demand for grid power. Smart grid-based regulators are more controllable than far more numerous consumer devices.

In another approach, a group of 16 western utilities called the Western Electric Industry Leaders called for mandatory use of "smart inverters". These devices convert DC to household AC and can also help with power quality. Such devices could eliminate the need for expensive utility equipment upgrades at a much lower total cost.

REFERENCES

1. Thompson, M.T. "Notes 01" (PDF). Introduction to Power Electronics. Thompson Consulting, Inc.
2. Kharagpur. "Power Semiconductor Devices" (PDF). EE IIT. Retrieved 25 March 2012.
3. Muhammad H. Rashid,POWER ELECTRONICS HANDBOOK DEVICES, CIRCUITS, AND APPLICATIONS Third Edition The structure introduced in this work is a multilevel inverter, which uses Separate DC Sources. The multilevel inverter using cascaded-inverter with SDCS synthesizes a desired voltage from several independent sources of DC voltages, which may be obtained from batteries, fuel cells, or solar cells. This configuration recently becomes very popular in AC power supply and adjustable speed drive applications. This new inverter can avoid extra clamping diodes or voltage balancing capacitors. Butterworth-Heinemann,2007 ISBN 978-0-12-382036-5

4. Hart, D. (2010). Power Electronics. McGraw-Hill Education. pp. Chapter 1. ISBN 978-0-07-128930-6.
5. Mohan, N. (2003). Power Electronics Converters Applications and Design. Michigan: John Wiley and Sons. pp. Chapter 1. ISBN 978-0-471-22693-2.
6. Bose, B (April 1992). "Evaluation of Modern Power Semiconductor Devices and Future Trends of Converters". IEEE Transactions on Industry Applications. 28 (2).
7. "semiconductor GTO". GTO. ABB. Retrieved 21 March 2012.
8. Rashid, M.H. (2001). Power Electronics Handbook. Academic Press. pp. 225–250.
9. Trzynadlowski, A.M. (2010). Introduction to Modern Power Electronics. Wiley. pp. 269–341.
10. Rahsid, M.H. (2010). Power Electronics Handbook: Devices, Circuits, and Applications. Elsevier. pp. 147–564. ISBN 978-0-12-382036-5.
11. Skvarenina, T.L. (2002). The power electronics handbook Industrial electronics series. CRC Press. pp. 94–140. ISBN 978-0-8493-7336-7.
12. Rashid, M.H. (2005). Digital power electronics and applications Electronics & Electrical. Academic Press. ISBN 978-0-12-088757-6.
13. Tolbert, L.M. "CYCLOCONVERTERS". University of Tennessee. Retrieved 23 March 2012.
14. Klumpner, C. "Power Electronics 2". Retrieved 23 March 2012.
15. Vodovozov, V (2006). Electronic engineering. ISBN 978-9985-69-039-0.
16. Lipo; Kim, Sul (2000). "AC/AC Power Conversion Based on Matric Converter Topology with Unidirectional Switches". IEEE Transactions on Industry Applications. 36 (1): 139–145. doi:10.1109/28.821808.
17. Wheeler; Wijekoon, Klumpner (July 2008). "Implementation of a Hybrid AC/AC Direct Power Converter with Unity Voltage Transfer Ratio". IEEE Transactions on Power Electronics. 23 (4): 1918–1986. doi:10.1109/tpel.2008.924601.
18. Khader, S. "THE APPLICATION OF PSIM & MATLAB/ SIMULINK IN POWER ELECTRONICS COURSES" (PDF). Retrieved 25 March 2012.
19. Bose, Bimal K. (September–October 1993). "Power Electronics and Motion Control – Technology Status and Recent Trends". IEEE.
20. Bose, Bimal K. (February 2009). "Power Electronics and Motor Drives Recent Progress and Perspective". IEEE.
21. Yano, Masao; Shigery Abe; Eiichi Ohno (2004). "History of Power Electronics for Motor Drives in Japan". IEEE.
22. D. J. Hammerstrom; et al. "Pacific Northwest GridWise™ Testbed Demonstration Projects, Part I. Olympic Peninsula Project" (PDF). Retrieved 2014-01-15.
23. U.S. Department of Energy. "Smart Grid / Department of Energy". Retrieved 2012-06-18.

24. Carrasco, Juan Manuel; Leopoldo Garcia Franquelo; Jan T. Bialasiewecz; Eduardo Galvan; Ramon C. Portillo Guisado; Ma. Angeles Martin Prats; Jose Ignacio Leon; Narciso Moreno-Alfonso (August 2006). "Power-Electronic Systems for the Grid Integration of Renewable Sources: A Survey". IEEE. 53 (4): 1002. doi:10.1109/tie.2006.878356.
25. LaMonica, Martin (2014-01-21). "Power Electronics Could Help Grid and Solar Power Get Along | MIT Technology Review". Technologyreview.com. Retrieved 2014-01-22.
26. Issa Batarseh, "Power Electronic Circuits" by John Wiley, 2003.
27. S.K. Mazumder, "High-Frequency Inverters: From Photovoltaic, Wind, and Fuel-Cell based Renewable- and Alternative-Energy DER/DG Systems to Battery based Energy-Storage Applications", Book Chapter in Power Electronics handbook, Editor M.H. Rashid, Academic Press, Burlington, Massachusetts, 2010.
28. V. Gureich "Electronic Devices on Discrete Components for Industrial and Power Engineering", CRC Press, New York, 2008, 418 p.
29. Editor: Semikron, Authors: Dr. Ulrich Nicolai, Dr. Tobias Reimann, Prof. Jürgen Petzoldt, Josef Lutz: Application Manual IGBT- and MOSFET-power modules, 1. edition, ISLE Verlag, 1998, ISBN 3-932633-24-5 online version
30. R. W. Erickson, D. Maksimovic, Fundamentals of Power Electronics, 2nd Ed., Springer, 2001, ISBN 0-7923-7270-0
31. Arendt Wintrich; Ulrich Nicolai; Werner Tursky; Tobias Reimann (2010), Applikationshandbuch 2010 (PDF-Version) (in German) (2. ed.), ISLE Verlag, ISBN 978-3-938843-56-7
32. Arendt Wintrich; Ulrich Nicolai; Werner Tursky; Tobias Reimann (2011), Application Manual 2011 (PDF-Version) (in German) (2. ed.), ISLE Verlag, ISBN 978-3-938843-66-6

CHAPTER 2

Power Semiconductor Devices

Chapter Contents

- **Introduction**
- **History**
- **Common Devices**
- **Classifications**
- **Parameters**
- **Research and Development**
- **Bipolar Junction Transistor (BJT)**
- **Mosfet**
- **Bootstrapping**
- **Dimmer**
- **Gate Turn-Off Thyristor**
- **Integrated Gate-Commutated Thyristor**

Power semiconductor devices are semiconductor devices used as switches or rectifiers in power electronic circuits (switch mode power supplies for example). They are also called power devices or when used in integrated circuits, called power ICs.

Some common power devices are the power diode, thyristor, power MOSFET and IGBT (insulated gate bipolar transistor). A power diode or MOSFET, for example, operates on similar principles as its low-power counterpart, but is able to carry a larger amount of current and typically is able to support a larger reverse-bias voltage in the off-state.

INTRODUCTION

A power semiconductor device is a semiconductor device used as a switch or rectifier in power electronics; a switch-mode power supply is an example. Such a device is also called a power device or, when used in an integrated circuit, a power IC.

A power semiconductor device is usually used in "commutation mode" (i.e., it is either on or off), and therefore has a design optimized for such usage; it should usually not be used in linear operation.

HISTORY

The first power semiconductor device appeared in 1952 with the introduction of the power diode by R.N. Hall. It was made of germanium and had a reverse voltage blocking capability of 200 V and a current rating of 35 A.

The thyristor appeared in 1957. It is able to withstand very high reverse breakdown voltage and is also capable of carrying high current. However, one disadvantage of the thyristor in switching circuits is that once it becomes 'latched-on' in the conducting state; it cannot be turned off by external control, as the thyristor turn-off is passive, i.e., the power must be disconnected from the device. Thyristors which could be turned off, called gate turn-off thyristors (GTO), were introduced in 1960. These overcome some limitations of the ordinary thyristor, because they can be turned on or off with an applied signal.

The first bipolar transistor device with substantial power handling capabilities was introduced in the 1948 by William Shockley. Due to improvements in the MOSFET technology (metal oxide semiconductor technology, initially developed to produce integrated circuits), the power MOSFET became available in the late 1970s. International Rectifier introduced a 25 A, 400 V power MOSFET in 1978. This device allows operation at higher frequencies than a bipolar transistor, but is limited to low voltage applications.

The Insulated-gate bipolar transistor (IGBT) was developed in the 1980s, and became widely available in the 1990s. This component has the power handling capability of the bipolar transistor and the advantages of the isolated gate drive of the power MOSFET.

COMMON DEVICES

Some common power devices are the power diode, thyristor, power MOSFET, and IGBT. The power diode and power MOSFET operate on similar principles to their low-power counterparts, but are able to carry a larger amount of current and are typically able to support a larger reverse-bias voltage in the *off-state*.

Structural changes are often made in a power device in order to accommodate the higher current density, higher power dissipation, and/or higher reverse breakdown voltage. The vast majority of the discrete (i.e., non-integrated) power devices are built using a vertical structure, whereas small-signal devices employ a lateral structure. With the vertical structure, the current rating of the device is proportional to its area, and the voltage blocking capability is achieved in the height of the die. With this structure, one of the connections of the device is located on the bottom of the semiconductor die.

CLASSIFICATIONS

A power device may be classified as one of the following main categories:

- A two-terminal device (e.g., a diode), whose state is completely dependent on the external power circuit to which it is connected.
- A three-terminal device (e.g., a triode), whose state is dependent on not only its external power circuit, but also the signal on its driving terminal (this terminal is known as the *gate* or *base*).

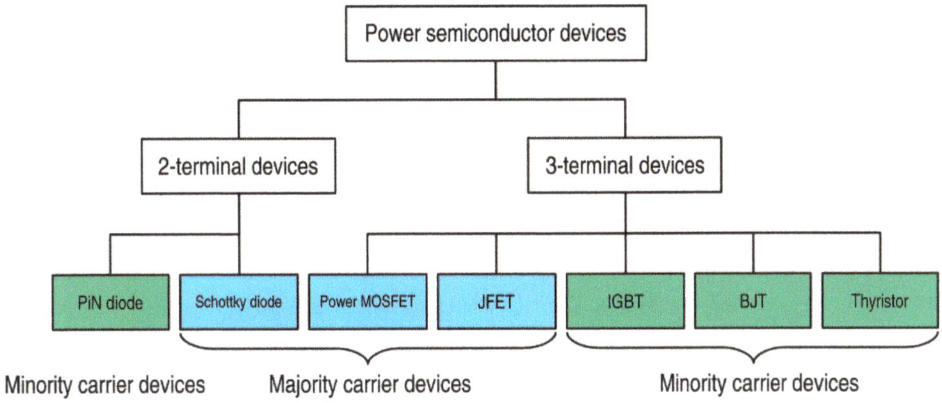

The power devices family, showing the principal power switches.

Another classification is less obvious, but has a strong influence on device performance:

- A *majority carrier device* (e.g., a Schottky diode, a MOSFET, etc.); this uses only one type of charge carriers.
- A *minority carrier device* (e.g., a thyristor, a bipolar transistor, an IGBT, etc.); this uses both majority and minority carriers (i.e., electrons and electron holes).

A majority carrier device is faster, but the charge injection of minority carrier devices allows for better on-state performance.

Diodes

An ideal diode should have the following characteristics:

- When *forward-biased*, the voltage across the end terminals of the diode should be zero, whatever the current that flows through it (on-state).
- When *reverse-biased*, the leakage current should be zero, whatever the voltage (off-state).
- The transition (or commutation) between the on-state and the off-state should be instantaneous.

In reality, the design of a diode is a trade-off between performance in on-state, off-state, and commutation. Indeed, the same area of the device must sustain the blocking voltage in the off-state and allow current flow in the on-state; as the requirements for the two states are completely opposite, a diode has to be either optimised for one of them, or time must be allowed to switch from one state to the other (i.e., the commutation speed must be reduced).

These trade-offs are the same for all power devices; for instance, a Schottky diode has excellent switching speed and on-state performance, but a high level of leakage current in the off-state. On the other hand, a PIN diode is commercially available in different commutation speeds (what are called "fast" and "ultrafast" rectifiers), but any increase in speed is necessarily associated with a lower performance in the on-state.

Switches

The trade-offs between voltage, current, and frequency ratings also exist for a switch. In fact, any power semiconductor relies on a PIN diode structure in order to sustain voltage; this can be seen in figure 2. The power MOSFET has the advantages of a majority carrier device, so it can achieve a very high operating frequency, but it cannot be used with high voltages; as it is a physical limit, no improvement is expected in the design of a silicon MOSFET concerning its maximum voltage ratings. However, its excellent performance in low voltage applications make it the device of choice (actually the only choice, currently) for applications with voltages below 200 V. By placing several devices in parallel, it is possible to increase the current rating of a switch. The MOSFET is particularly suited to this configuration, because its positive thermal coefficient of resistance tends to result in a balance of current between the individual devices.

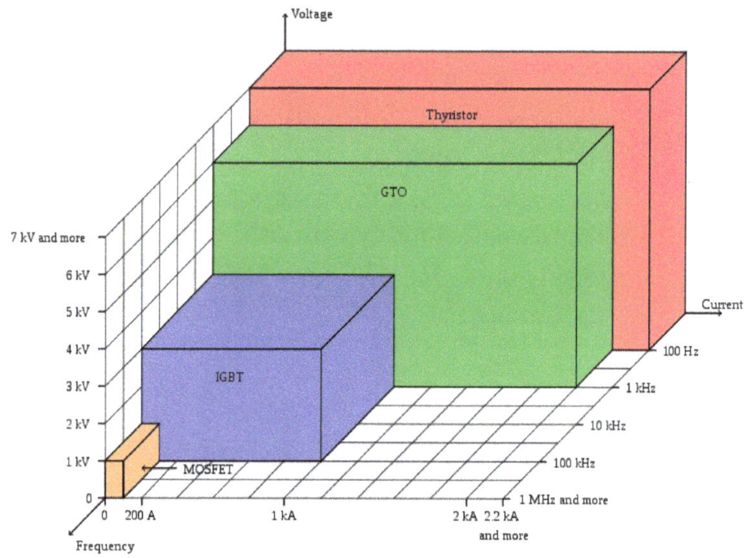

Current/Voltage/switching frequency domains of the main power electronics switches.

The IGBT is a recent component, so its performance improves regularly as technology evolves. It has already completely replaced the bipolar transistor in power applications; a power module is available in which several IGBT devices are connected in parallel, making it attractive for power levels up to several megawatts, which pushes further the limit at which thyristors and GTOs become the only option. Basically, an IGBT is a bipolar transistor driven by a power MOSFET; it has the advantages of being a minority carrier device (good performance in the on-state, even for high voltage devices), with the high input impedance of a MOSFET (it can be driven on or off with a very low amount of power).

The major limitation of the IGBT for low voltage applications is the high voltage drop it exhibits in the on-state (2-to-4 V). Compared to the MOSFET, the operating frequency of the IGBT is relatively low (usually not higher than 50 kHz), mainly because of a problem during turn-off known as *current-tail*: The slow decay of the conduction current during turn-off results from a slow recombination of a large number of carriers that flood the thick 'drift' region of the IGBT during conduction. The net result is that the turn-off switching loss of an IGBT is considerably higher than its turn-on loss. Generally, in datasheets, turn-off energy is mentioned as a measured parameter; that number has to be multiplied with the switching frequency of the intended application in order to estimate the turn-off loss.

At very high power levels, a thyristor-based device (e.g., a SCR, a GTO, a MCT, etc.) is still the only choice. This device can be turned on by a pulse provided by a driving circuit, but cannot be turned off by removing the pulse. A thyristor turns off as soon as no more current flows through it; this happens automatically in an alternating current system on each cycle, or requires a circuit with the means to divert current around the device. Both MCTs and GTOs have been developed to overcome this limitation, and are widely used in power distribution applications.

PARAMETERS

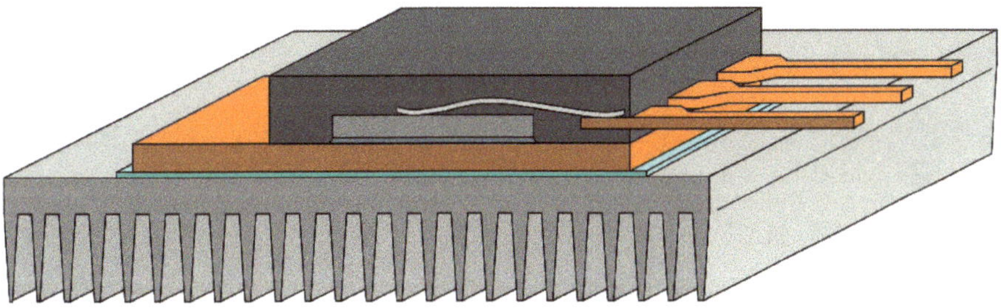

A power device is usually attached to a heatsink to remove the heat caused by operation losses.

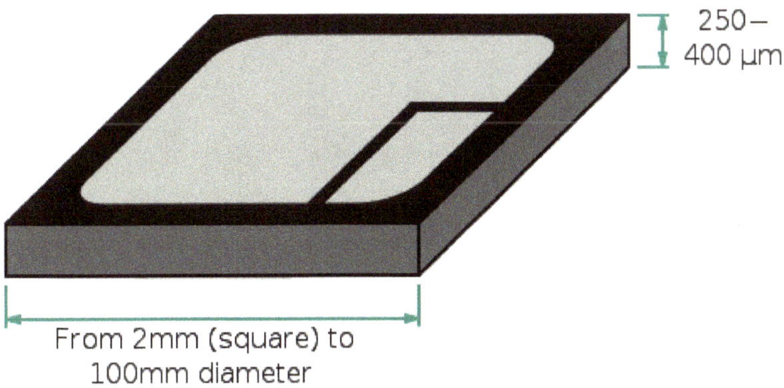

The power semiconductor die of a three-terminal device (IGBT, MOSFET or BJT). Two contacts are on top of the die, the remaining one is on the back.

- Breakdown voltage: Often, there is a trade-off between breakdown voltage rating and on-resistance, because increasing the breakdown voltage by incorporating a thicker and lower doped drift region leads to a higher on-resistance.
- On-resistance: A higher current rating lowers the on-resistance due to greater numbers of parallel cells. This increases overall capacitance and slows down the speed.
- Rise and fall times: The amount of time it takes to switch between the on-state and the off-state.
- Safe-operating area: This is a thermal dissipation and "latch-up" consideration.
- Thermal resistance: This is an often ignored but extremely important parameter from the point of view of practical design; a semiconductor does not perform well at elevated temperature, and yet due to large current conduction, a power semiconductor device invariably heats up. Therefore, such a devices needs to be cooled by removing that heat continuously; packaging and heatsink technology provide a means for removing heat from a semiconductor device by conducting it to the external environment. Generally, a large current device has a large die and packaging surface areas and lower thermal resistance.

RESEARCH AND DEVELOPMENT

Packaging

The role of packaging is to:

- connect a die to the external circuit.
- provide a way to remove the heat generated by the device.
- protect the die from the external environment (moisture, dust, etc.).

Many of the reliability issues of a power device are either related to excessive temperature or fatigue due to thermal cycling. Research is currently carried out on the following topics:

- Cooling performance.
- Resistance to thermal cycling by closely matching the Coefficient of thermal expansion of the packaging to that of the silicon.
- The maximum operating temperature of the packaging material.

Research is also ongoing on electrical issues such as reducing the parasitic inductance of packaging; this inductance limits the operating frequency, because it generates losses during commutation.

A low-voltage MOSFET is also limited by the parasitic resistance of its package, as its intrinsic on-state resistance can be as low as one or two milliohms.

Some of the most common type of power semiconductor packages include the TO-220, TO-247, TO-262, TO-3, D^2Pak, etc.

Improvement of Structures

The IGBT design is still under development and can be expected to provide increases in operating voltages. At the high-power end of the range, the MOS-controlled thyristor is a promising device. Achieving a major improvement over the conventional MOSFET structure by employing the super junction charge-balance principle: essentially, it allows the thick drift region of a power MOSFET to be heavily doped, thereby reducing the electrical resistance to electron flow without compromising the breakdown voltage. This is juxtaposed with a region that is similarly doped with the opposite carrier polarity (*holes*); these two similar, but oppositely doped regions effectively cancel out their mobile charge and develop a 'depleted region' that supports the high voltage during the off-state. On the other hand, during the on-state, the higher doping of the drift region allows for the easy flow of carriers, thereby reducing on-resistance. Commercial devices, based on this super junction principle, have been developed by companies like Infineon (CoolMOS™ products) and International Rectifier (IR).

Wide Band-Gap Semiconductors

The major breakthrough in power semiconductor devices is expected from the replacement of silicon by a wide band-gap semiconductor. At the moment, silicon carbide (SiC) is considered to be the most promising. A SiC Schottky diode with a breakdown voltage of 1200 V is commercially available, as is a 1200 V JFET. As both are majority carrier devices, they can operate at high speed. A bipolar device is being developed for higher voltages (up to 20 kV). Among its advantages, silicon carbide can operate at a higher temperature (up to 400 °C) and has a lower thermal resistance than silicon, allowing for better cooling.

BIPOLAR JUNCTION TRANSISTOR (BJT)

A bipolar junction transistor (bipolar transistor or BJT) is a type of transistor that uses both electron and hole charge carriers. In contrast, unipolar transistors, such as field-effect transistors, only use one kind of charge carrier. For their operation, BJTs use two junctions between two semiconductor types, n-type and p-type.

BJTs are manufactured in two types, NPN and PNP, and are available as individual components, or fabricated in integrated circuits, often in large numbers. The basic function of a BJT is to amplify current. This allows BJTs to be used as amplifiers or switches, giving them wide applicability in electronic equipment, including computers, televisions, mobile phones, audio amplifiers, industrial control, and radio transmitters.

BJT schematic symbols

Note on Current Direction

By convention, the direction of current on diagrams is shown as the direction that a positive charge would move. This is called *conventional current*. However, current in metal conductors is due to the flow of electrons which, because they carry a negative charge, move in the opposite direction to conventional current. On the other hand, inside a bipolar transistor, currents can be composed of both positively charged holes and negatively charged electrons. In this article, current arrows are shown in the conventional direction, but labels for the movement of holes and electrons show their actual direction inside the transistor.

Function

BJTs come in two types, or polarities, known as PNP and NPN based on the doping types of the three main terminal regions. An NPN transistor comprises two semiconductor junctions that share a thin p-doped anode region, and a PNP transistor comprises two semiconductor junctions that share a thin n-doped cathode region.

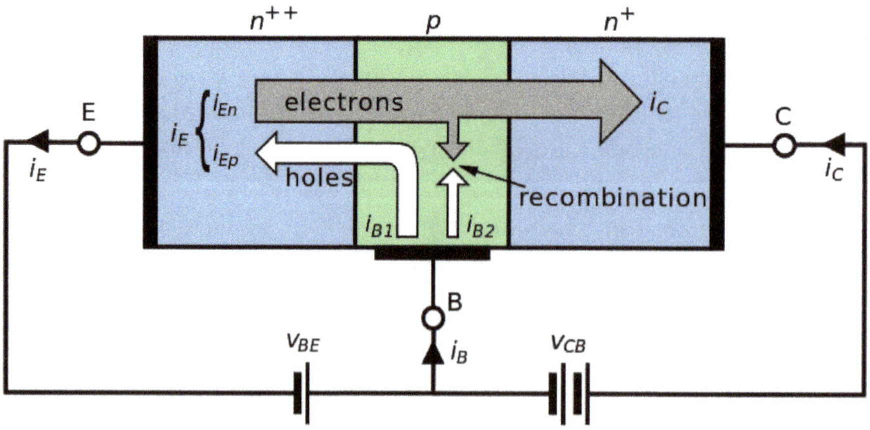

NPN BJT with forward-biased E–B junction and reverse-biased B–C junction

Charge flow in a BJT is due to diffusion of charge carriers across a junction between two regions of different charge concentrations. The regions of a BJT are called *emitter*, *collector*, and *base*. A discrete transistor has three leads for connection to these regions. Typically, the emitter region is heavily doped compared to the other two layers, whereas the majority charge carrier concentrations in base and collector layers are about the same. By design, most of the BJT collector current is due to the flow of charges injected from a high-concentration emitter into the base where they are minority carriers that diffuse toward the collector, and so BJTs are classified as minority-carrier devices.

In typical operation, the base–emitter junction is forward biased, which means that the p-doped side of the junction is at a more positive potential than the n-doped side, and the base–collector junction is reverse biased. In an NPN transistor, when positive bias is applied to the base–emitter junction, the equilibrium is disturbed between the thermally generated carriers and the repelling electric field of the n-doped emitter depletion region. This allows thermally excited electrons to inject from the emitter into the base region. These electrons diffuse through the base from the region of high concentration near the emitter towards the region of low concentration near the collector. The electrons in the base are called *minority carriers* because the base is doped p-type, which makes holes the *majority carrier* in the base.

To minimize the percentage of carriers that recombine before reaching the collector–base junction, the transistor's base region must be thin enough that carriers can diffuse across it in much less time than the semiconductor's minority carrier lifetime. In particular, the thickness of the base must be much less than the diffusion length of the electrons. The collector–base junction is reverse-biased, and so little electron injection occurs from the collector to the base, but electrons that diffuse through the base towards the collector are swept into the collector by the electric field in the depletion region of the collector–base junction. The thin *shared* base and asymmetric collector–emitter doping are what differentiates a bipolar transistor from two *separate* and oppositely biased diodes connected in series.

Voltage, Current, and Charge Control

The collector–emitter current can be viewed as being controlled by the base–emitter current (current control), or by the base–emitter voltage (voltage control). These views are related by the current–voltage relation of the base–emitter junction, which is just the usual exponential current–voltage curve of a p-n junction (diode).

The physical explanation for collector current is the concentration of minority carriers in the base region. Due to low level injection (in which there are much fewer excess carriers than normal majority carriers) the ambipolar transport rates (in which the excess majority and minority carriers flow at the same rate) is in effect determined by the excess minority carriers.

Detailed transistor models of transistor action, such as the Gummel–Poon model, account for the distribution of this charge explicitly to explain transistor behavior more exactly. The charge-control view easily handles phototransistors, where minority carriers in the base region are created by the absorption of photons, and handles the dynamics of turn-off, or recovery time, which depends on charge in the base region recombining. However, because base charge is not a signal that is visible at the terminals, the current- and voltage-control views are generally used in circuit design and analysis.

In analog circuit design, the current-control view is sometimes used because it is approximately linear. That is, the collector current is approximately β_F times the base current. Some basic circuits can be designed by assuming that the emitter–base voltage is approximately constant, and that collector current is beta times the base current. However, to accurately and reliably design production BJT circuits, the voltage-control (for example, Ebers–Moll) model is required. The voltage-control model requires an exponential function to be taken into account, but when it is linearized such that the transistor can be modeled as a transconductance, as in the Ebers–Moll model, design for circuits such as differential amplifiers again becomes a mostly linear problem, so the voltage-control view is often preferred. For translinear circuits, in which the exponential I–V curve is key to the operation, the transistors are usually modeled as voltage-controlled current sources whose transconductance is proportional to their collector current. In general, transistor-level circuit design is performed using SPICE or a comparable analog circuit simulator, so model complexity is usually not of much concern to the designer.

Turn-On, Turn-Off, and Storage Delay

The bipolar transistor exhibits a few delay characteristics when turning on and off. Most transistors, and especially power transistors, exhibit long base-storage times that limit maximum frequency of operation in switching applications. One method for reducing this storage time is by using a Baker clamp.

Transistor Parameters: Alpha (α) and Beta (β)

The proportion of electrons able to cross the base and reach the collector is a measure of the BJT efficiency. The heavy doping of the emitter region and light doping of the base region causes many more electrons to be injected from the emitter into the base than holes to be injected from the base into the emitter.

The *common-emitter current gain* is represented by β_F or the h-parameter h_{FE}; it is approximately the ratio of the DC collector current to the DC base current in forward-active region. It is typically greater than 50 for small-signal transistors but can be smaller in transistors designed for high-power applications.

Another important parameter is the *common-base current gain*, α_F. The common-base current gain is approximately the gain of current from emitter to collector in the forward-active region. This ratio usually has a value close to unity; between 0.980 and 0.998. It is less than unity due to recombination of charge carriers as they cross the base region.

Alpha and beta are more precisely related by the following identities (NPN transistor):

$$\alpha_F = \frac{I_{\mathrm{C}}}{I_{\mathrm{E}}}, \qquad \beta_F = \frac{I_{\mathrm{C}}}{I_{\mathrm{B}}}$$

$$\alpha_F = \frac{\beta_F}{1+\beta_F} \iff \beta_F = \frac{\alpha_F}{1-\alpha_F}$$

Structure

A BJT consists of three differently doped semiconductor regions: the *emitter* region, the *base* region and the *collector* region. These regions are, respectively, *p* type, *n* type and *p* type in a PNP transistor, and *n* type, *p* type and *n* type in an NPN transistor. Each semiconductor region is connected to a terminal, appropriately labeled: *emitter* (E), *base* (B) and *collector* (C).

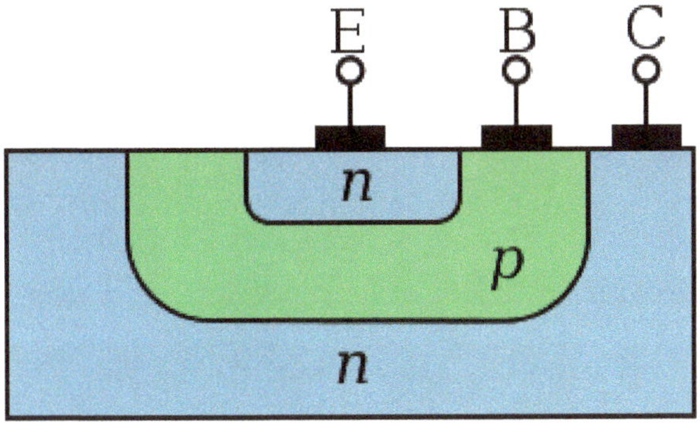

Simplified cross section of a planar *NPN* bipolar junction transistor

The *base* is physically located between the *emitter* and the *collector* and is made from lightly doped, high-resistivity material. The collector surrounds the emitter region, making it almost impossible for the electrons injected into the base region to escape without being collected, thus making the resulting value of α very close to unity, and so, giving the transistor a large β. A cross-section view of a BJT indicates that the collector–base junction has a much larger area than the emitter–base junction.

The bipolar junction transistor, unlike other transistors, is usually not a symmetrical device. This means that interchanging the collector and the emitter makes the transistor leave the forward active mode and start to operate in reverse mode. Because the transistor›s internal structure is usually optimized for forward-mode operation, interchanging the collector and the emitter makes the values of α and β in reverse operation much smaller than those in forward operation; often the α of the reverse mode is lower than 0.5. The lack of symmetry is primarily due to the doping ratios of the emitter and the collector. The emitter is heavily doped, while the collector is lightly doped, allowing a large reverse bias voltage to be applied before the collector–base junction breaks down. The collector–base junction is reverse biased in normal operation. The reason the emitter is heavily doped is to increase the emitter injection efficiency: the ratio of carriers injected by the emitter to those injected by the base. For high current gain, most of the carriers injected into the emitter–base junction must come from the emitter.

Die of a KSY34 high-frequency NPN transistor. Bond wires connect to the base and emitter

The low-performance "lateral" bipolar transistors sometimes used in CMOS processes are sometimes designed symmetrically, that is, with no difference between forward and backward operation.

Small changes in the voltage applied across the base–emitter terminals cause the current between the *emitter* and the *collector* to change significantly. This effect can be used to amplify the input voltage or current. BJTs can be thought of as voltage-controlled current sources, but are more simply characterized as current-controlled current sources, or current amplifiers, due to the low impedance at the base.

Early transistors were made from germanium but most modern BJTs are made from silicon. A significant minority are also now made from gallium arsenide, especially for very high speed applications.

NPN

NPN is one of the two types of bipolar transistors, consisting of a layer of P-doped semiconductor (the "base") between two N-doped layers. A small current entering the base is amplified to produce a large collector and emitter current. That is, when there is a positive potential difference measured from the base of an NPN transistor to its emitter (that is, when the base is **high** relative to the emitter), as well as a positive potential difference measured from the collector to the emitter, the transistor becomes active. In this "on" state, charge flows from the collector to the emitter of the transistor. Most of the current is carried by electrons moving from emitter to collector as minority carriers in the P-type base region. To allow for greater current and faster operation, most bipolar transistors used today are NPN because electron mobility is higher than hole mobility.

The symbol of an NPN BJT. A mnemonic for the symbol is "**n**ot **p**ointing i**n**".

A mnemonic device for the NPN transistor symbol is "*n*ot *p*ointing i*n*", based on the arrows in the symbol and the letters in the name.

PNP

The other type of BJT is the PNP, consisting of a layer of N-doped semiconductor between two layers of P-doped material. A small current leaving the base is amplified in the collector output. That is, a PNP transistor is "on" when its base is pulled **low** relative to the emitter. In a PNP transistor, the emitter–base region is forward biased, so holes are injected into the base as minority carriers. The base is very thin, and most of the holes cross the reverse-biased base–collector junction to the collector.

The symbol of a PNP BJT. A mnemonic for the symbol is "**p**oints **in** **p**roudly".

The arrows in the NPN and PNP transistor symbols are on the emitter legs and point in the direction of the conventional current when the device is in forward active or forward saturated mode.

A mnemonic device for the PNP transistor symbol is "*p*ointing i*n* (*p*roudly/*p*ermanently)", based on the arrows in the symbol and the letters in the name.

Heterojunction Bipolar Transistor

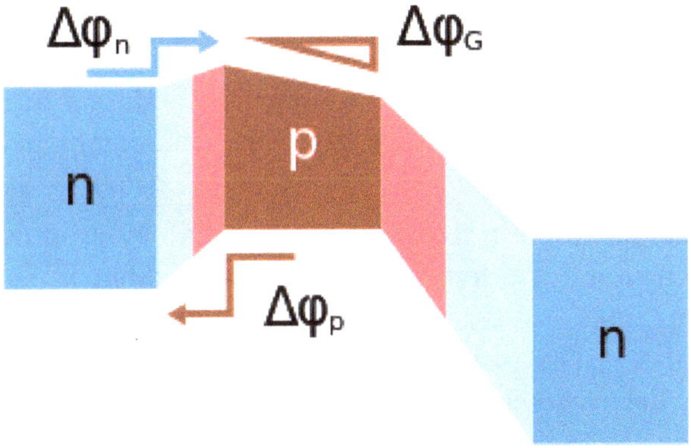

Bands in graded heterojunction NPN bipolar transistor. Barriers indicated for electrons to move from emitter to base, and for holes to be injected backward from base to emitter; also, grading of bandgap in base assists electron transport in base region; light colors indicate depleted regions.

The heterojunction bipolar transistor (HBT) is an improvement of the BJT that can handle signals of very high frequencies up to several hundred GHz. It is common in modern ultrafast circuits, mostly RF systems.

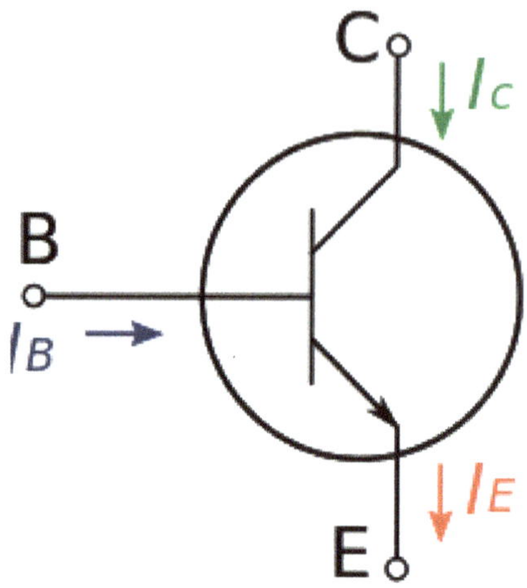

Symbol for NPN Bipolar Transistor with current flow direction.

Heterojunction transistors have different semiconductors for the elements of the transistor. Usually the emitter is composed of a larger bandgap material than the base. The figure shows that this difference in bandgap allows the barrier for holes to inject backward from the base into the emitter, denoted in the figure as $\Delta\varphi_p$, to be made large, while the barrier for electrons to inject into the base $\Delta\varphi_n$ is made low. This barrier arrangement helps reduce minority carrier injection from the base when the emitter-base junction is under forward bias, and thus reduces base current and increases emitter injection efficiency.

The improved injection of carriers into the base allows the base to have a higher doping level, resulting in lower resistance to access the base electrode. In the more traditional BJT, also referred to as homojunction BJT, the efficiency of carrier injection from the emitter to the base is primarily determined by the doping ratio between the emitter and base, which means the base must be lightly doped to obtain high injection efficiency, making its resistance relatively high. In addition, higher doping in the base can improve figures of merit like the Early voltage by lessening base narrowing.

The grading of composition in the base, for example, by progressively increasing the amount of germanium in a SiGe transistor, causes a gradient in bandgap in the neutral base, denoted in the figure by $\Delta\varphi_G$, providing a "built-in" field that assists electron transport across the base. That drift component of transport aids the normal diffusive transport, increasing the frequency response of the transistor by shortening the transit time across the base.

Two commonly used HBTs are silicon–germanium and aluminum gallium arsenide, though a wide variety of semiconductors may be used for the HBT structure. HBT structures are usually grown by epitaxy techniques like MOCVD and MBE.

MOSFET

The metal–oxide–semiconductor field-effect transistor (MOSFET, MOS-FET, or MOS FET) is a type of field-effect transistor (FET). It has an insulated gate, whose voltage determines the conductivity of the device. This ability to change conductivity with the amount of applied voltage can be used for amplifying or switching electronic signals. Although FET is sometimes used when referring to MOSFET devices, other types of field-effect transistors also exist.

Although the MOSFET is a four-terminal device with source (S), gate (G), drain (D), and body (B) terminals, the body (or substrate) of the MOSFET is often connected to the source terminal, making it a three-terminal device like other field-effect transistors. Because these two terminals are normally connected to each other (short-circuited) internally, only three terminals appear in electrical diagrams.

The basic principle of the field-effect transistor was first patented by Julius Edgar Lilienfeld in 1925.

The main advantage of a MOSFET over a regular transistor is that it requires very little current to turn on (less than 1mA), while delivering a much higher current to a load (10 to 50A or more).

In *enhancement mode* MOSFETs, a voltage drop across the oxide induces a conducting channel between the source and drain contacts *via* the field effect. The term "enhancement mode" refers to the increase of conductivity with increase in oxide field that adds carriers to the channel, also referred to as the *inversion layer*. The channel can contain electrons (called an nMOSFET or nMOS), or holes (called a pMOSFET or pMOS), opposite in type to the substrate, so nMOS is made with a p-type substrate, and pMOS with an n-type substrate. In the less common *depletion mode* MOSFET, detailed later on, the channel consists of carriers in a surface impurity layer of opposite type to the substrate, and conductivity is decreased by application of a field that depletes carriers from this surface layer.

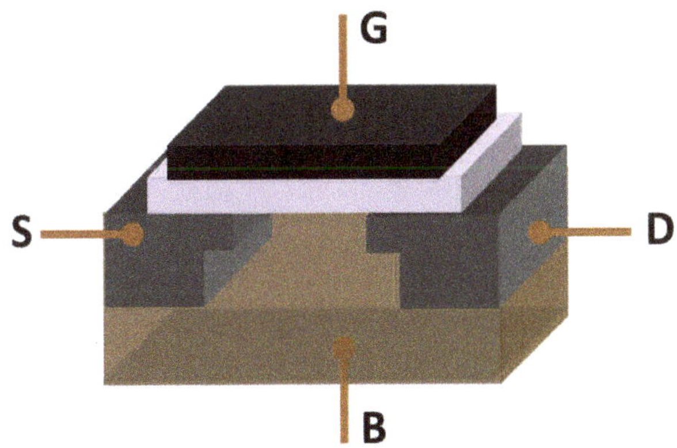

MOSFET showing gate (G), body (B), source (S) and drain (D) terminals. The gate is separated from the body by an insulating layer (white)

Two power MOSFETs in D2PAK surface-mount packages. Operating as switches, each of these components can sustain a blocking voltage of 120 volts in the OFF state, and can conduct a continuous current of 30 amperes in the ON state, dissipating up to about 100 watts and controlling a load of over 2000 watts. A matchstick is pictured for scale

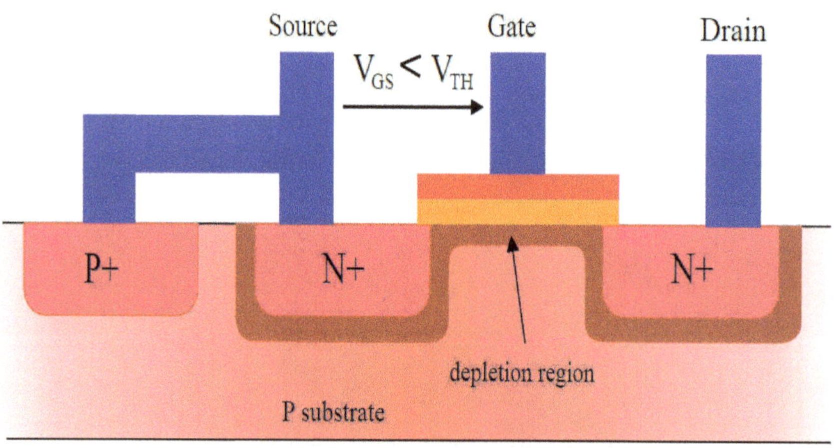

A cross section through an nMOSFET when the gate voltage VGS is below the threshold for making a conductive channel; there is little or no conduction between the terminals drain and source; the switch is off. When the gate is more positive, it attracts electrons, inducing an n-type conductive channel in the substrate below the oxide, which allows electrons to flow between the n-doped terminals; the switch is on

The "metal" in the name MOSFET is now often a misnomer because the previously metal gate material is now often a layer of polysilicon (polycrystalline silicon). Aluminium had been the gate material until the mid-1970s, when polysilicon became dominant, due to its capability to form self-aligned gates. Metallic gates are regaining popularity, since it is difficult to increase the speed of operation of transistors without metal gates. Likewise, the "oxide" in the name can be a misnomer, as different dielectric materials are used with the aim of obtaining strong channels with smaller applied voltages.

An insulated-gate field-effect transistor or IGFET is a related term almost synonymous with MOSFET. The term may be more inclusive, since many "MOSFETs" use a gate that is not metal, and a gate insulator that is not oxide. Another synonym is MISFET for metal–insulator–semiconductor FET.

The MOSFET is by far the most common transistor in both digital and analog circuits, though the bipolar junction transistor was at one time much more common.

History

The basic principle of this kind of transistor was first patented by Julius Edgar Lilienfeld in 1925. Twenty five years later, when Bell Telephone attempted to patent the junction transistor, they found Lilienfeld already holding a patent, worded in a way that would include all types of transistors. Bell Labs was able to work out an agreement with Lilienfeld, who was still alive at that time (it is not known if they paid him money or not). It was at that time the Bell Labs version was given the name bipolar junction transistor, or simply junction transistor, and Lilienfeld's design took the name field effect transistor.

In 1959, Dawon Kahng and Martin M. (John) Atalla at Bell Labs invented the metal–oxide–semiconductor field-effect transistor (MOSFET) as an offshoot to the patented FET design.Operationally and structurally different from the bipolar junction transistor, the MOSFET was made by putting an insulating layer on the surface of the semiconductor and then placing a metallic gate electrode on that. It used crystalline silicon for the semiconductor and a thermally oxidized layer of silicon dioxide for the insulator. The silicon MOSFET did not generate localized electron traps at the interface between the silicon and its native oxide layer, and thus was inherently free from the trapping and scattering of carriers that had impeded the performance of earlier field-effect transistors.

Discrete power MOSFETs are currently widely used as low voltage switches.

Composition

Usually the semiconductor of choice is silicon, but some chip manufacturers, most notably IBM and Intel, recently started using a chemical compound of silicon and germanium (SiGe) in MOSFET channels. Unfortunately, many semiconductors with better electrical properties than silicon, such as gallium arsenide, do not form good semiconductor-to-insulator interfaces, and thus are not suitable for MOSFETs. Research continues on creating insulators with acceptable electrical characteristics on other semiconductor material.

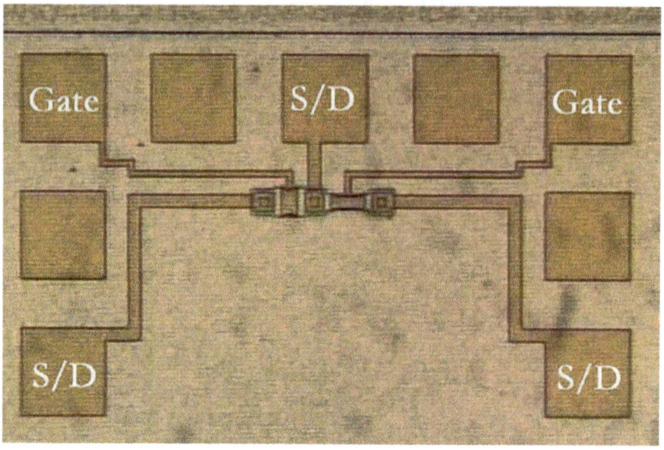

Photomicrograph of two metal-gate MOSFETs in a test pattern. Probe pads for two gates and three source/drain nodes are labeled.

In order to overcome the increase in power consumption due to gate current leakage, a high-κ dielectric is used instead of silicon dioxide for the gate insulator, while polysilicon is replaced by metal gates.

The gate is separated from the channel by a thin insulating layer, traditionally of silicon dioxide and later of silicon oxynitride. Some companies have started to introduce a high-κ dielectric + metal gate combination in the 45 nanometer node.

When a voltage is applied between the gate and body terminals, the electric field generated penetrates through the oxide and creates an "inversion layer" or "channel" at the semiconductor-insulator interface. The inversion channel is of the same type, p-type or n-type, as the source and drain, and thus it provides a channel through which current can pass. Varying the voltage between the gate and body modulates the conductivity of this layer and thereby controls the current flow between drain and source. This is known as enhancement mode.

Operation

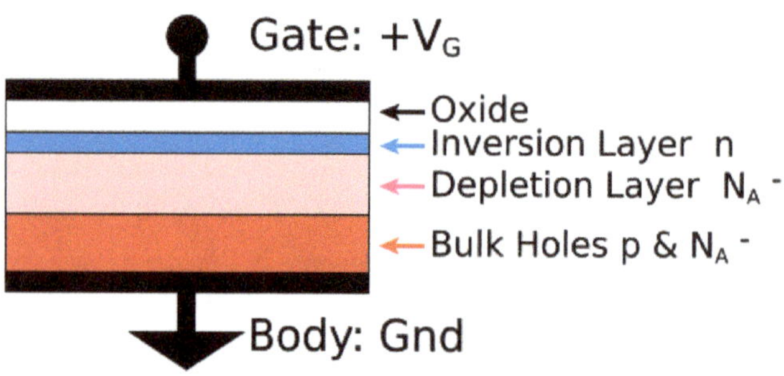

Metal–oxide–semiconductor structure on p-type silicon.

Metal–Oxide–Semiconductor Structure

The traditional metal–oxide–semiconductor (MOS) structure is obtained by growing a layer of silicon dioxide (SiO_2) on top of a silicon substrate and depositing a layer of metal or polycrystalline silicon (the latter is commonly used). As the silicon dioxide is a dielectric material, its structure is equivalent to a planar capacitor, with one of the electrodes replaced by a semiconductor.

When a voltage is applied across a MOS structure, it modifies the distribution of charges in the semiconductor. If we consider a p-type semiconductor (with N_A the density of acceptors, p the density of holes; $p = N_A$ in neutral bulk), a positive voltage, V_{GB}, from gate to body creates a depletion layer by forcing the positively charged holes away from the gate-insulator/semiconductor interface, leaving exposed a carrier-free region of immobile, negatively charged acceptor ions. If V_{GB} is high enough, a high concentration of negative charge carriers forms in an *inversion layer* located in a thin layer next to the interface between the semiconductor and the insulator. Unlike the MOSFET, where the inversion layer electrons are supplied rapidly from the source/drain electrodes, in the MOS capacitor they are produced much more slowly by thermal generation through carrier generation and recombination centers in the depletion region. Conventionally, the gate voltage at which the volume density of electrons in the inversion layer is the same as the volume density of holes in the body is called the threshold voltage. When the voltage between transistor gate and source (V_{GS}) exceeds the threshold voltage (V_{th}), it is known as overdrive voltage.

This structure with p-type body is the basis of the n-type MOSFET, which requires the addition of n-type source and drain regions.

Structure and Channel Formation

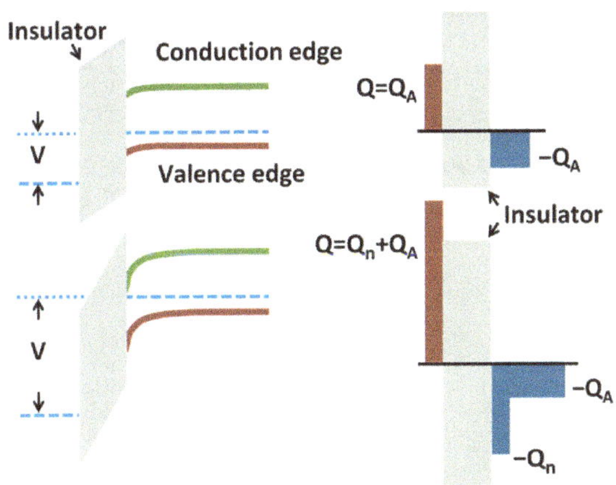

Channel formation in nMOS MOSFET shown as band diagram: Top panels: An applied gate voltage bends bands, depleting holes from surface (left). The charge inducing the bending is balanced by a layer of negative acceptor-ion charge (right). Bottom panel: A larger applied voltage further depletes holes but conduction band lowers enough in energy to populate a conducting channel

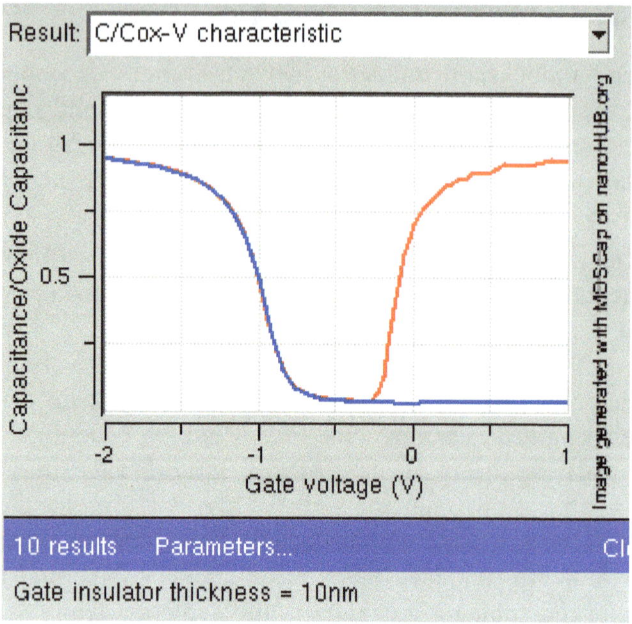

C–V profile for a bulk MOSFET with different oxide thickness. The leftmost part of the curve corresponds to accumulation. The valley in the middle corresponds to depletion. The curve on the right corresponds to inversion.

A MOSFET is based on the modulation of charge concentration by a MOS capacitance between a *body* electrode and a *gate* electrode located above the body and insulated from all other device regions by a gate dielectric layer which in the case of a MOSFET is an oxide, such as silicon dioxide. If dielectrics other than an oxide such as silicon dioxide (often referred to as oxide) are employed the device may be referred to as a metal–insulator–semiconductor FET (MISFET). Compared to the MOS capacitor, the MOSFET includes two additional terminals (*source* and *drain*), each connected to individual highly doped regions that are separated by the body region. These regions can be either p or n type, but they must both be of the same type, and of opposite type to the body region. The source and drain (unlike the body) are highly doped as signified by a "+" sign after the type of doping.

If the MOSFET is an n-channel or nMOS FET, then the source and drain are "n+" regions and the body is a "p" region. If the MOSFET is a p-channel or pMOS FET, then the source and drain are "p+" regions and the body is a "n" region. The source is so named because it is the source of the charge carriers (electrons for n-channel, holes for p-channel) that flow through the channel; similarly, the drain is where the charge carriers leave the channel.

The occupancy of the energy bands in a semiconductor is set by the position of the Fermi level relative to the semiconductor energy-band edges.

With sufficient gate voltage, the valence band edge is driven far from the Fermi level, and holes from the body are driven away from the gate.

At larger gate bias still, near the semiconductor surface the conduction band edge is brought close to the Fermi level, populating the surface with electrons in an *inversion layer* or *n-channel* at the interface between the p region and the oxide. This conducting channel extends between the source and the drain, and current is conducted through it when a voltage is applied between the two electrodes. Increasing the voltage on the gate leads to a higher electron density in the inversion layer and therefore increases the current flow between the source and drain. For gate voltages below the threshold value, the channel is lightly populated, and only a very small subthreshold leakage current can flow between the source and the drain.

When a negative gate-source voltage (positive source-gate) is applied, it creates a *p-channel* at the surface of the n region, analogous to the n-channel case, but with opposite polarities of charges and voltages. When a voltage less negative than the threshold value (a negative voltage for the p-channel) is applied between gate and source, the channel disappears and only a very small subthreshold current can flow between the source and the drain. The device may comprise a silicon on insulator (SOI) device in which a buried oxide (BOX) is formed below a thin semiconductor layer. If the channel region between the gate dielectric and a BOX region is very thin, the channel is referred to as an ultrathin channel (UTC) region with the source and drain regions formed on either side thereof in and/or above the thin semiconductor layer. Alternatively, the device may comprise a semiconductor on insulator (SEMOI) device in which semiconductors other than silicon are employed. Many alternative semiconductor materials may be employed. When the source and drain regions are formed above the channel in whole or in part, they are referred to as raised source/drain (RSD) regions.

Circuit Symbols

A variety of symbols are used for the MOSFET. The basic design is generally a line for the channel with the source and drain leaving it at right angles and then bending back at right angles into the same direction as the channel. Sometimes three line segments are used for enhancement mode and a solid line for depletion mode. Another line is drawn parallel to the channel for the gate.

The "bulk" or "body" connection, if shown, is shown connected to the back of the channel with an arrow indicating pMOS or nMOS. Arrows always point from P to N, so an NMOS (N-channel in P-well or P-substrate) has the arrow pointing in (from the bulk to the channel). If the bulk is connected to the source (as is generally the case with discrete devices) it is sometimes angled to meet up with the source leaving the transistor. If the bulk is not shown (as is often the case in IC design as they are generally common bulk) an inversion symbol is sometimes used to indicate PMOS, alternatively an arrow on the source may be used in the same way as for bipolar transistors (out for nMOS, in for pMOS).

Comparison of enhancement-mode and depletion-mode MOSFET symbols, along with JFET symbols. The orientation of the symbols, (most significantly the position of source relative to drain) is such that more positive voltages appear higher on the page than less positive voltages, implying current flowing "down" the page:

Comparison of n- and p-type MOSFETs		
Parameter	**nMOSFET**	**pMOSFET**
Source/drain type	n-type	p-type
Channel type (MOS capacitor)	n-type	p-type
Gate type (poly Si)	n+ poly-Si	p+ poly-Si
Gate type (metal)	φ_m ~ Si CB	φ_m ~ Si VB
Well type	p-type	n-type
Threshold voltage, V_{th}	positive (enhancement) negative (depletion)	negative (enhancement) positive (depletion)
Band-bending	Downwards	Upwards
Inversion layer carriers	electrons	holes
Substrate type	p-type	n-type

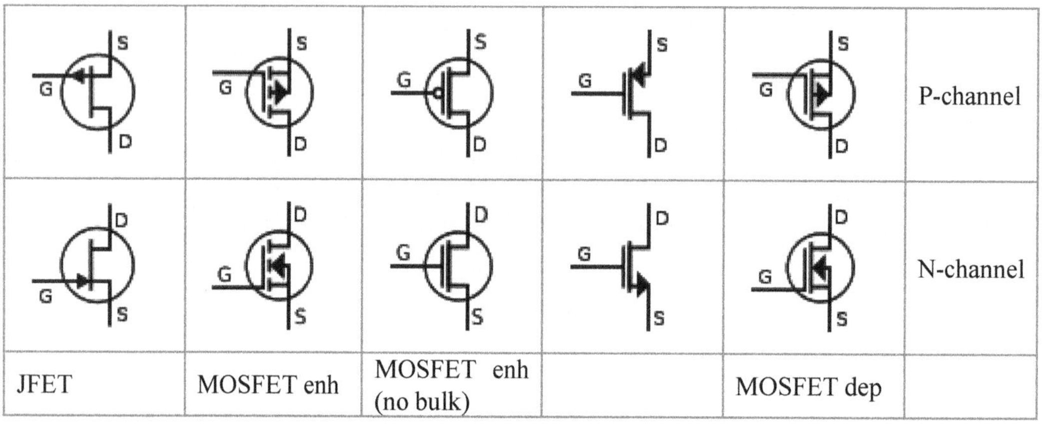

In schematics where G, S, D are not labeled, the detailed features of the symbol indicate which terminal is source and which is drain. For enhancement-mode and depletion-mode MOSFET symbols (in columns two and five), the source terminal is the one connected to the triangle. Additionally, in this diagram, the gate is shown as an "L" shape, whose input leg is closer to S than D, also indicating which is which. However, these symbols are often drawn with a "T" shaped gate (as elsewhere on this page), so it is the triangle which must be relied upon to indicate the source terminal.

For the symbols in which the bulk, or body, terminal is shown, it is here shown internally connected to the source (i.e., the black triangles in the diagrams in columns 2 and 5). This is a typical configuration, but by no means the only important configuration. In general, the MOSFET is a four-terminal device, and in integrated circuits many of the MOSFETs share a body connection, not necessarily connected to the source terminals of all the transistors.

Applications

Digital integrated circuits such as microprocessors and memory devices contain thousands to millions of integrated MOSFET transistors on each device, providing the basic switching functions required to implement logic gates and data storage. Discrete devices are widely used in applications such as switch mode power supplies, variable-frequency drives and other power electronics applications where each device may be switching hundreds or thousands of watts. Radio-frequency amplifiers up to the UHF spectrum use MOSFET transistors as analog signal and power amplifiers. Radio systems also use MOSFETs as oscillators, or mixers to convert frequencies. MOSFET devices are also applied in audio-frequency power amplifiers for public address systems, sound reinforcement and home and automobile sound systems

MOS Integrated Circuits

Following the development of clean rooms to reduce contamination to levels never before thought necessary, and of photolithography and the planar process to allow circuits to be made in very few steps, the Si–SiO_2 system possessed such technical attractions as low cost of production (on a per circuit basis) and ease of integration. Largely because of these two factors, the MOSFET has become the most widely used type of transistor in integrated circuits.

General Microelectronics introduced the first commercial MOS integrated circuit in 1964.

Additionally, the method of coupling two complementary MOSFETS (P-channel and N-channel) into one high/low switch, known as CMOS, means that digital circuits dissipate very little power except when actually switched.

The earliest microprocessors starting in 1970 were all "MOS microprocessors"—i.e., fabricated entirely from PMOS logic or fabricated entirely from NMOS logic. In the 1970s, "MOS microprocessors" were often contrasted with "CMOS microprocessors" and "bipolar bit-slice processors".

CMOS Circuits

The MOSFET is used in digital complementary metal–oxide–semiconductor (CMOS) logic, which uses p- and n-channel MOSFETs as building blocks. Overheating is a major concern in integrated circuits since ever more transistors are packed into ever smaller chips. CMOS logic reduces power consumption because no current flows (ideally), and thus no power is consumed, except when the inputs to logic gates are being switched. CMOS accomplishes this current reduction by complementing every nMOSFET with a pMOSFET and connecting both gates and both drains together. A high voltage on the gates will cause the nMOSFET to conduct and the pMOSFET not to conduct and a low voltage on the gates causes the reverse. During the switching time as the voltage goes from one state to another, both MOSFETs will conduct briefly. This arrangement greatly reduces power consumption and heat generation.

Digital

The growth of digital technologies like the microprocessor has provided the motivation to advance MOSFET technology faster than any other type of silicon-based transistor. A big advantage of MOSFETs for digital switching is that the oxide layer between the gate and the channel prevents DC current from flowing through the gate, further reducing power consumption and giving a very large input impedance. The insulating oxide between the gate and channel effectively isolates a MOSFET in one logic stage from earlier and later stages, which allows a single MOSFET output to drive a considerable number of MOSFET inputs. Bipolar transistor-based logic (such as TTL) does not have such a high fanout capacity. This isolation also makes it easier for the designers to ignore to some extent loading effects between logic stages independently. That extent is defined by the operating frequency: as frequencies increase, the input impedance of the MOSFETs decreases.

Analog

The MOSFET's advantages in digital circuits do not translate into supremacy in all analog circuits. The two types of circuit draw upon different features of transistor behavior. Digital circuits switch, spending most of their time outside the switching region, while analog circuits depend on the linearity of response when the MOSFET is held precisely in the switching region. The bipolar junction transistor (BJT) has traditionally been the analog designer's transistor of choice, due largely to its higher transconductance and its lower output impedance (drain-voltage independence) in the switching region.

Nevertheless, MOSFETs are widely used in many types of analog circuits because of certain advantages. The characteristics and performance of many analog circuits can be scaled up or down by changing the sizes (length and width) of the MOSFETs used. By comparison, in most bipolar transistors the size of the device does not significantly affect its performance. MOSFETs' ideal characteristics regarding gate current (zero) and drain-source offset voltage (zero) also make them nearly ideal switch elements, and also make switched capacitor analog circuits practical. In their linear region, MOSFETs can be used as precision resistors, which can have a much higher controlled resistance than BJTs. In high power circuits, MOSFETs sometimes have the advantage of not suffering from thermal runaway as BJTs do. Also, MOSFETs can be configured to perform as capacitors and gyrator circuits which allow op-amps made from them to appear as inductors, thereby allowing all of the normal analog devices on a chip (except for diodes, which can be made smaller than a MOSFET anyway) to be built entirely out of MOSFETs. This means that complete analog circuits can be made on a silicon chip in a much smaller space and with simpler fabrication techniques. MOSFETS are ideally suited to switch inductive loads because of tolerance to inductive kickback.

Some ICs combine analog and digital MOSFET circuitry on a single mixed-signal integrated circuit, making the needed board space even smaller. This creates a need to isolate the analog circuits from the digital circuits on a chip level, leading to the use of isolation rings and Silicon-On-Insulator (SOI). Since MOSFETs require more space to handle a given amount of power than a BJT, fabrication processes can incorporate BJTs and MOSFETs into a single device. Mixed-transistor devices are called Bi-FETs (bipolar FETs) if they contain just one BJT-FET and BiCMOS (bipolar-CMOS) if they contain complementary BJT-FETs. Such devices have the advantages of both insulated gates and higher current density.

Analog Switches

MOSFET analog switches use the MOSFET to pass analog signals when on, and as a high impedance when off. Signals flow in both directions across a MOSFET switch. In this application, the drain and source of a MOSFET exchange places depending on the relative voltages of the source/drain electrodes. The source is the more negative side for an N-MOS or the more positive side for a P-MOS. All of these switches are limited on what signals they can pass or stop by their gate–source, gate–drain and source–drain voltages; exceeding the voltage, current, or power limits will potentially damage the switch.

Single-Type

This analog switch uses a four-terminal simple MOSFET of either P or N type.

In the case of an n-type switch, the body is connected to the most negative supply (usually GND) and the gate is used as the switch control. Whenever the gate voltage exceeds the source voltage by at least a threshold voltage, the MOSFET conducts. The higher the voltage, the more the MOSFET can conduct. An N-MOS switch passes all voltages less than $V_{gate}–V_{tn}$. When the switch is conducting, it typically operates in the linear (or ohmic) mode of operation, since the source and drain voltages will typically be nearly equal.

In the case of a P-MOS, the body is connected to the most positive voltage, and the gate is brought to a lower potential to turn the switch on. The P-MOS switch passes all voltages higher than $V_{gate}–V_{tp}$ (threshold voltage V_{tp} is negative in the case of enhancement-mode P-MOS).

Dual-Type (CMOS)

This "complementary" or CMOS type of switch uses one P-MOS and one N-MOS FET to counteract the limitations of the single-type switch. The FETs have their drains and sources connected in parallel, the body of the P-MOS is connected to the high potential (V_{DD}) and the body of the N-MOS is connected to the low potential (Gnd). To turn the switch on, the gate of the P-MOS is driven to the low potential and the gate of the N-MOS is driven to the high potential. For voltages between $V_{DD}–V_{tn}$ and Gnd–V_{tp}, both FETs conduct the signal; for voltages less than Gnd–V_{tp}, the N-MOS conducts alone; and for voltages greater than $V_{DD}–V_{tn}$, the P-MOS conducts alone.

The voltage limits for this switch are the gate–source, gate–drain and source–drain voltage limits for both FETs. Also, the P-MOS is typically two to three times wider than the N-MOS, so the switch will be balanced for speed in the two directions.

Tri-state circuitry sometimes incorporates a CMOS MOSFET switch on its output to provide for a low-ohmic, full-range output when on, and a high-ohmic, mid-level signal when off.

Construction

Gate Material

The primary criterion for the gate material is that it is a good conductor. Highly doped polycrystalline silicon is an acceptable but certainly not ideal conductor, and also suffers from some more technical deficiencies in its role as the standard gate material. Nevertheless, there are several reasons favoring use of polysilicon:

- The threshold voltage (and consequently the drain to source on-current) is modified by the work function difference between the gate material and channel material. Because polysilicon is a semiconductor, its work function can be modulated by adjusting the type and level of doping. Furthermore, because polysilicon has the same bandgap as the underlying silicon channel, it is quite straightforward to tune the work function to achieve low threshold voltages for both NMOS and PMOS devices. By contrast, the work functions of metals are not easily modulated, so tuning the work function to obtain low threshold voltages becomes a significant challenge. Additionally, obtaining low-threshold devices on both PMOS and NMOS devices sometimes requires the use of different metals for each device type. While bimetallic integrated circuits (i.e., one type of metal for gate electrodes of NFETS and a second type of metal for gate electrodes of PFETS) are not common, they are known in patent literature and provide some benefit in terms of tuning electrical circuits' overall electrical perforamnce.
- The silicon-SiO_2 interface has been well studied and is known to have relatively few defects. By contrast many metal–insulator interfaces contain significant levels of defects which can lead to Fermi level pinning, charging, or other phenomena that ultimately degrade device performance.
- In the MOSFET IC fabrication process, it is preferable to deposit the gate material prior to certain high-temperature steps in order to make better-performing transistors. Such high temperature steps would melt some metals, limiting the types of metal that can be used in a metal-gate-based process.

While polysilicon gates have been the de facto standard for the last twenty years, they do have some disadvantages which have led to their likely future replacement by metal gates. These disadvantages include:

- Polysilicon is not a great conductor (approximately 1000 times more resistive than metals) which reduces the signal propagation speed through the material. The resistivity can be lowered by increasing the level of doping, but even highly doped polysilicon is not as conductive as most metals. To improve conductivity further, sometimes a high-temperature metal such as tungsten, titanium, cobalt, and more recently nickel is alloyed with the top layers of the polysilicon. Such a blended material is called silicide. The silicide-polysilicon combination has better electrical properties than polysilicon alone and still does not melt in subsequent processing. Also the threshold voltage is not significantly higher than with polysilicon alone, because the silicide material is not near the channel. The process in which silicide is formed on both the gate electrode and the source and drain regions is sometimes called salicide, self-aligned silicide.

- When the transistors are extremely scaled down, it is necessary to make the gate dielectric layer very thin, around 1 nm in state-of-the-art technologies. A phenomenon observed here is the so-called poly depletion, where a depletion layer is formed in the gate polysilicon layer next to the gate dielectric when the transistor is in the inversion. To avoid this problem, a metal gate is desired. A variety of metal gates such as tantalum, tungsten, tantalum nitride, and titanium nitride are used, usually in conjunction with high-k dielectrics. An alternative is to use fully silicided polysilicon gates, a process known as FUSI.

Present high performance CPUs use metal gate technology, together with high-k dielectrics, a combination known as HKMG (High-K, Metal Gate). The disadvantages of metal gates are overcome by a few techniques:

- The threshold voltage is tuned by including a thin "work function metal" layer between the high-K dielectric and the main metal. This layer is thin enough that the total work function of the gate is influenced by both the main metal and thin metal work functions (either due to alloying during annealing, or simply due to the incomplete screening by the thin metal). The threshold voltage thus can be tuned by the thickness of the thin metal layer.
- High-K dielectrics are now well studied, and their defects are understood.
- HKMG processes exist that do not require the metals to experience high temperature anneals; other processes select metals that can survive the annealing step.

Insulator

As devices are made smaller, insulating layers are made thinner, and at some point tunneling of carriers through the insulator from the channel to the gate electrode takes place. To reduce the resulting leakage current, the insulator can be made thicker by choosing a material with a higher dielectric constant. To see how thickness and dielectric constant are related, note that Gauss's law connects field to charge as:

$$Q = \kappa\epsilon_0 E,$$

with Q = charge density, κ = dielectric constant, ε_0 = permittivity of empty space and E = electric field. From this law it appears the same charge can be maintained in the channel at a lower field provided κ is increased. The voltage on the gate is given by:

$$V_G = V_{ch} + E\, t_{ins} = V_{ch} + \frac{Q t_{ins}}{\kappa\epsilon_0},$$

with V_G = gate voltage, V_{ch} = voltage at channel side of insulator, and t_{ins} = insulator thickness. This equation shows the gate voltage will not increase when the insulator thickness increases, provided κ increases to keep t_{ins}/κ = *constant*.

The insulator in a MOSFET is a dielectric which can in any event be silicon oxide, but many other dielectric materials are employed. The generic term for the dielectric is gate dielectric since the dielectric lies directly below the gate electrode and above the channel of the MOSFET.

Junction Design

The source-to-body and drain-to-body junctions are the object of much attention because of three major factors: their design affects the current-voltage (*I-V*) characteristics of the device, lowering output resistance, and also the speed of the device through the loading effect of the junction capacitances, and finally, the component of stand-by power dissipation due to junction leakage.

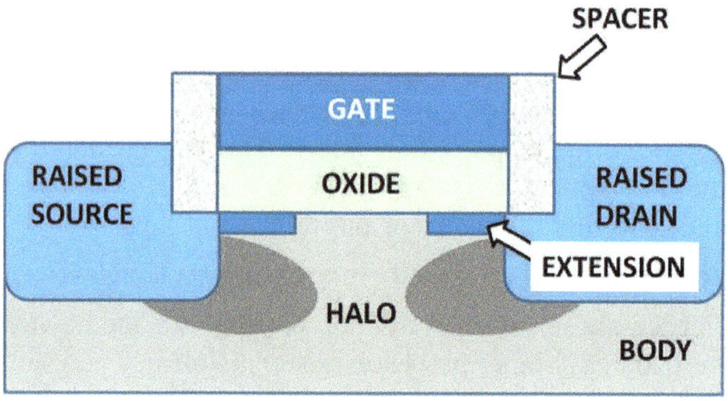

MOSFET showing shallow junction extensions, raised source and drain and halo implant. Raised source and drain separated from gate by oxide spacers

The drain induced barrier lowering of the threshold voltage and channel length modulation effects upon *I-V* curves are reduced by using shallow junction extensions. In addition, *halo* doping can be used, that is, the addition of very thin heavily doped regions of the same doping type as the body tight against the junction walls to limit the extent of depletion regions.

The capacitive effects are limited by using raised source and drain geometries that make most of the contact area border thick dielectric instead of silicon.

These various features of junction design are shown (with artistic license) in the figure.

Scaling

Over the past decades, the MOSFET has continually been scaled down in size; typical MOSFET channel lengths were once several micrometres, but modern integrated circuits are incorporating MOSFETs with channel lengths of tens of nanometers. Robert Dennard's work on scaling theory was pivotal in recognizing that this ongoing reduction was possible. Intel began production of a process featuring a 32 nm feature size (with the channel being even shorter) in late 2009. The semiconductor industry maintains a "roadmap", the ITRS, which sets the pace for MOSFET development. Historically, the difficulties with decreasing the size of the MOSFET have been associated with the semiconductor device fabrication process, the need to use very low voltages, and with poorer electrical performance necessitating circuit redesign and innovation (small MOSFETs exhibit higher leakage currents and lower output resistance).

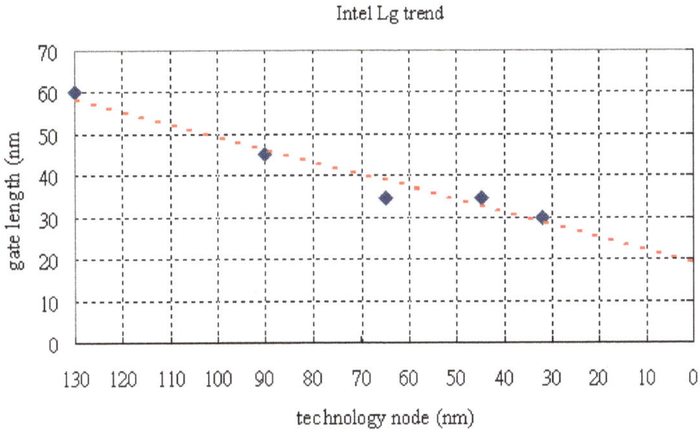

Trend of Intel CPU transistor gate length

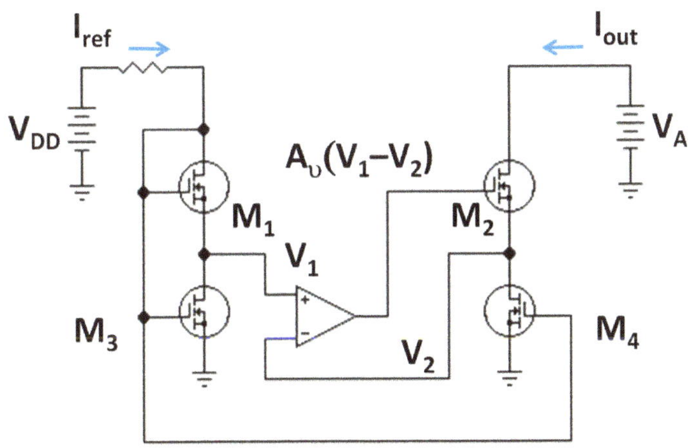

MOSFET version of gain-boosted current mirror; M_1 and M_2 are in active mode, while M_3 and M_4 are in Ohmic mode, and act like resistors. The operational amplifier provides feedback that maintains a high output resistance

Smaller MOSFETs are desirable for several reasons. The main reason to make transistors smaller is to pack more and more devices in a given chip area. This results in a chip with the same functionality in a smaller area, or chips with more functionality in the same area. Since fabrication costs for a semiconductor wafer are relatively fixed, the cost per integrated circuits is mainly related to the number of chips that can be produced per wafer. Hence, smaller ICs allow more chips per wafer, reducing the price per chip. In fact, over the past 30 years the number of transistors per chip has been doubled every 2–3 years once a new technology node is introduced. For example, the number of MOSFETs in a microprocessor fabricated in a 45 nm technology can well be twice as many as in a 65 nm chip. This doubling of transistor density was first observed by Gordon Moore in 1965 and is commonly referred to as Moore's law. It is also expected that smaller transistors switch faster. For example, one approach to size reduction is a scaling of the MOSFET that requires all device dimensions to reduce proportionally. The main device dimensions are the channel length, channel width, and oxide thickness. When they are scaled down by equal factors, the transistor channel resistance does not change, while gate capacitance is cut by that factor. Hence, the RC delay of the transistor scales with a similar factor. While this has been traditionally the case for the older technologies, for the state-of-the-art MOSFETs reduction of the transistor dimensions does not necessarily translate to higher chip speed because the delay due to interconnections is more significant.

Producing MOSFETs with channel lengths much smaller than a micrometre is a challenge, and the difficulties of semiconductor device fabrication are always a limiting factor in advancing integrated circuit technology. Though processes such as ALD have improved fabrication for small components, the small size of the MOSFET (less than a few tens of nanometers) has created operational problems:

Higher Subthreshold Conduction

As MOSFET geometries shrink, the voltage that can be applied to the gate must be reduced to maintain reliability. To maintain performance, the threshold voltage of the MOSFET has to be reduced as well. As threshold voltage is reduced, the transistor cannot be switched from complete turn-off to complete turn-on with the limited voltage swing available; the circuit design is a compromise between strong current in the "on" case and low current in the "off" case, and the application determines whether to favor one over the other. Subthreshold leakage (including subthreshold conduction, gate-oxide leakage and reverse-biased junction leakage), which was ignored in the past, now can consume upwards of half of the total power consumption of modern high-performance VLSI chips.

Increased Gate-Oxide Leakage

The gate oxide, which serves as insulator between the gate and channel, should be made as thin as possible to increase the channel conductivity and performance when the transistor is on and to reduce subthreshold leakage when the transistor is off. However, with current gate oxides with a thickness of around 1.2 nm (which in silicon is ~5 atoms thick) the quantum mechanical phenomenon of electron tunneling occurs between the gate and channel, leading to increased power consumption. Silicon dioxide has traditionally been used as the gate insulator. Silicon dioxide however has a modest dielectric constant. Increasing the dielectric constant of the gate dielectric allows a thicker layer while maintaining a high capacitance (capacitance is proportional to dielectric constant and inversely proportional to dielectric thickness). All else equal, a higher dielectric thickness reduces the quantum tunneling current through the dielectric between the gate and the channel. Insulators that have a larger dielectric constant than silicon dioxide (referred to as high-k dielectrics), such as group IVb metal silicates e.g. hafnium and zirconium silicates and oxides are being used to reduce the gate leakage from the 45 nanometer technology node onwards. On the other hand, the barrier height of the new gate insulator is an important consideration; the difference in conduction band energy between the semiconductor and the dielectric (and the corresponding difference in valence band energy) also affects leakage current level. For the traditional gate oxide, silicon dioxide, the former barrier is approximately 8 eV. For many alternative dielectrics the value is significantly lower, tending to increase the tunneling current, somewhat negating the advantage of higher dielectric constant. The maximum gate-source voltage is determined by the strength of the electric field able to be sustained by the gate dielectric before significant leakage occurs. As the insulating dielectric is made thinner, the electric field strength within it goes up for a fixed voltage. This necessitates using lower voltages with the thinner dielectric.

Increased Junction Leakage

To make devices smaller, junction design has become more complex, leading to higher doping levels, shallower junctions, "halo" doping and so forth, all to decrease drain-induced barrier lowering. To keep these complex junctions in place, the annealing steps formerly used to remove damage and electrically active defects must be curtailed increasing junction leakage. Heavier doping is also associated with thinner depletion layers and more recombination centers that result in increased leakage current, even without lattice damage.

DIBL and V_T Roll Off

Because of the short-channel effect, channel formation is not entirely done by the gate, but now the drain and source also affect the channel formation. As the channel length decreases, the depletion regions of the source and drain come closer together and make the threshold voltage (V_T) a function of the length of the channel. This is called V_T roll-off. V_T also becomes function of drain to source voltage V_{DS}. As we increase the V_{DS}, the depletion regions increase in size, and a considerable amount of charge is depleted by the V_{DS}. The gate voltage required to form the channel is then lowered, and thus, the V_T decreases with an increase in V_{DS}. This effect is called drain induced barrier lowering (DIBL).

Lower Output Resistance

For analog operation, good gain requires a high MOSFET output impedance, which is to say, the MOSFET current should vary only slightly with the applied drain-to-source voltage. As devices are made smaller, the influence of the drain competes more successfully with that of the gate due to the growing proximity of these two electrodes, increasing the sensitivity of the MOSFET current to the drain voltage. To counteract the resulting decrease in output resistance, circuits are made more complex, either by requiring more devices, for example the cascode and cascade amplifiers, or by feedback circuitry using operational amplifiers, for example a circuit like that in the adjacent figure.

Lower Transconductance

The transconductance of the MOSFET decides its gain and is proportional to hole or electron mobility (depending on device type), at least for low drain voltages. As MOSFET size is reduced, the fields in the channel increase and the dopant impurity levels increase. Both changes reduce the carrier mobility, and hence the transconductance. As channel lengths are reduced without proportional reduction in drain voltage, raising the electric field in the channel, the result is velocity saturation of the carriers, limiting the current and the transconductance.

Interconnect Capacitance

Traditionally, switching time was roughly proportional to the gate capacitance of gates. However, with transistors becoming smaller and more transistors being placed on the chip, interconnect capacitance (the capacitance of the metal-layer connections between different parts of the chip) is becoming a large percentage of capacitance. Signals have to travel through the interconnect, which leads to increased delay and lower performance.

Heat Production

The ever-increasing density of MOSFETs on an integrated circuit creates problems of substantial localized heat generation that can impair circuit operation. Circuits operate more slowly at high temperatures, and have reduced reliability and shorter lifetimes. Heat sinks and other cooling devices and methods are now required for many integrated circuits including microprocessors. Power MOSFETs are at risk of thermal runaway. As their on-state resistance rises with temperature, if the load is approximately a constant-current load then the power loss rises correspondingly, generating further heat. When the heatsink is not able to keep the temperature low enough, the junction temperature may rise quickly and uncontrollably, resulting in destruction of the device.

Process Variations

With MOSFETs becoming smaller, the number of atoms in the silicon that produce many of the transistor's properties is becoming fewer, with the result that control of dopant numbers and placement is more erratic. During chip manufacturing, random process variations affect all transistor dimensions: length, width, junction depths, oxide thickness *etc.*, and become a greater percentage of overall transistor size as the transistor shrinks. The transistor characteristics become less certain, more statistical. The random nature of manufacture means we do not know which particular example MOSFETs actually will end up in a particular instance of the circuit. This uncertainty forces a less optimal design because the design must work for a great variety of possible component MOSFETs.

Modeling Challenges

Modern ICs are computer-simulated with the goal of obtaining working circuits from the very first manufactured lot. As devices are miniaturized, the complexity of the processing makes it difficult to predict exactly what the final devices look like, and modeling of physical processes becomes more challenging as well. In addition, microscopic variations in structure due simply to the probabilistic nature of atomic processes require statistical (not just deterministic) predictions. These factors combine to make adequate simulation and "right the first time" manufacture difficult.

Other Types

Dual-Gate

The dual-gate MOSFET has a tetrode configuration, where both gates control the current in the device. It is commonly used for small-signal devices in radio frequency applications where biasing the drain-side gate at constant potential reduces the gain loss caused by Miller effect, replacing two separate transistors in cascode configuration. Other common uses in RF circuits include gain control and mixing (frequency conversion). The "tetrode" description, though accurate, does not replicate the vacuum-tube tetrode. Vacuum-tube tetrodes, using a screen grid, exhibit much lower grid-plate capacitance and much higher output impedance and voltage gains than triode vacuum tubes. These improvements are commonly an order of magnitude (10 times) or considerably more. Tetrode transistors (whether bipolar junction or field-effect) do not exhibit improvements of such a great degree.

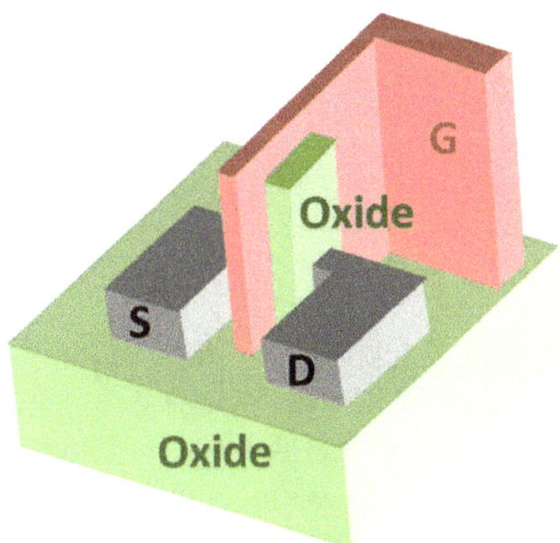

A FinFET

The FinFET is a double-gate silicon-on-insulator device, one of a number of geometries being introduced to mitigate the effects of short channels and reduce drain-induced barrier lowering. The "fin" refers to the narrow channel between source and drain. A thin insulating oxide layer on either side of the fin separates it from the gate. SOI FinFETs with a thick oxide on top of the fin are called *double-gate* and those with a thin oxide on top as well as on the sides are called *triple-gate* FinFETs.

Depletion-Mode

There are *depletion-mode* MOSFET devices, which are less commonly used than the standard *enhancement-mode* devices already described. These are MOSFET devices that are doped so that a channel exists even with zero voltage from gate to source. To control the channel, a negative voltage is applied to the gate (for an n-channel device), depleting the channel, which reduces the current flow through the device. In essence, the depletion-mode device is equivalent to a normally closed (on) switch, while the enhancement-mode device is equivalent to a normally open (off) switch. Due to their low noise figure in the RF region, and better gain, these devices are often preferred to bipolars in RF front-ends such as in TV sets. Depletion-mode MOSFET families include BF 960 by Siemens and BF 980 by Philips (dated 1980s), whose derivatives are still used in AGC and RF mixer front-ends.

NMOS Logic

For devices of equal current driving capability, n-channel MOSFETs can be made smaller than p-channel MOSFETs, due to p-channel charge carriers (holes) having lower mobility than do n-channel charge carriers (electrons), and producing only one type of MOSFET on a silicon substrate is cheaper and technically simpler. These were the driving principles in the design of NMOS logic which uses n-channel MOSFETs exclusively. However, neglecting leakage current, unlike CMOS logic, NMOS logic consumes power even when no switching is taking place. With advances in technology, CMOS logic displaced NMOS logic in the mid-1980s to become the preferred process for digital chips.

Power MOSFET

Power MOSFETs have a different structure. As with most power devices, the structure is vertical and not planar. Using a vertical structure, it is possible for the transistor to sustain both high blocking voltage and high current. The voltage rating of the transistor is a function of the doping and thickness of the N-epitaxial layer, while the current rating is a function of the channel width (the wider the channel, the higher the current). In a planar structure, the current and breakdown voltage ratings are both a function of the channel dimensions (respectively width and length of the channel), resulting in inefficient use of the "silicon estate". With the vertical structure, the component area is roughly proportional to the current it can sustain, and the component thickness (actually the N-epitaxial layer thickness) is proportional to the breakdown voltage.

Power MOSFETs with lateral structure are mainly used in high-end audio amplifiers and high-power PA systems. Their advantage is a better behavior in the saturated region (corresponding to the linear region of a bipolar transistor) than the vertical MOSFETs. Vertical MOSFETs are designed for switching applications.

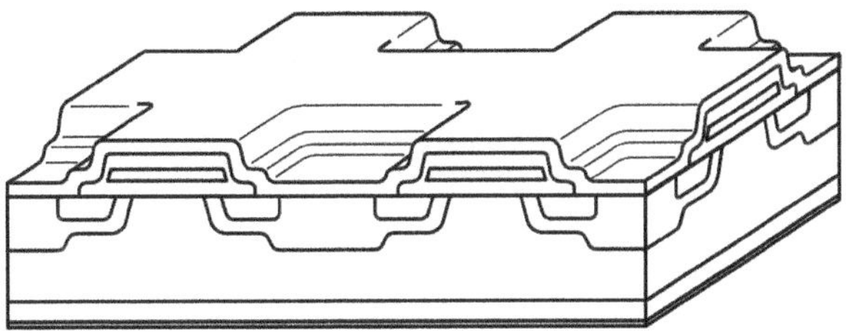

Cross section of a power MOSFET, with square cells. A typical transistor is constituted of several thousand cells

DMOS

DMOS stands for double-diffused metal–oxide–semiconductor. There are *LDMOS* (Lateral Double-diffused Metal Oxide Semiconductor) and *VDMOS* (Vertical Double-diffused Metal Oxide Semiconductor). Most power MOSFETs are made using this technology.

RHBD

Semiconductor sub-micrometer and nanometer electronic circuits are the primary concern for operating within the normal tolerance in harsh radiation environments like outer space. One of the design approaches for making a radiation-hardened-by-design (RHBD) device is Enclosed-Layout-Transistor (ELT). Normally, the gate of the MOSFET surrounds the drain, which is placed in the center of the ELT. The source of the MOSFET surrounds the gate. Another RHBD MOSFET is called H-Gate. Both of these transistors have very low leakage current with respect to radiation. However, they are large in size and take more space on silicon than a standard MOSFET. In older STI (shallow trench isolation) designs, radiation strikes near the silicon oxide region cause the channel inversion at the corners of the standard MOSFET due to accumulation of radiation induced trapped charges. If the charges are large enough, the accumulated charges affect STI surface edges along the channel near the channel interface (gate) of the standard MOSFET. Thus the device channel inversion occurs along the channel edges and the device creates off-state leakage path, causing device to turn on. So the reliability of circuits degrades severely. The ELT offers many advantages. These advantages include improvement of reliability by reducing unwanted surface inversion at the gate edges that occurs in the standard MOSFET. Since the gate edges are enclosed in ELT, there is no gate oxide edge (STI at gate interface), and thus the transistor off-state leakage is reduced very much. Low-power microelectronic circuits including computers, communication devices and monitoring systems in space shuttle and satellites are very different from what we use on earth. They are radiation (high-speed atomic particles like proton and neutron, solar flare magnetic energy dissipation in Earth's space, energetic cosmic rays like X-ray, gamma ray etc.) tolerant circuits. These special electronics are designed by applying very different techniques using RHBD MOSFETs to ensure the safe space journey and also space-walk of astronauts.

Floating-gate MOSFET

The floating-gate MOSFET (FGMOS) is a field-effect transistor, whose structure is similar to a conventional MOSFET. The gate of the FGMOS is electrically isolated, creating a floating node in DC, and a number of secondary gates or inputs are deposited above the floating gate (FG) and are electrically isolated from it. These inputs are only capacitively connected to the FG. Since the FG is completely surrounded by highly resistive material, the charge contained in it remains unchanged for long periods of time. Usually Fowler-Nordheim tunneling and hot-carrier injection mechanisms are used to modify the amount of charge stored in the FG.

Some applications of the FGMOS are digital storage element in EPROM, EEPROM and flash memories, neuronal computational element in neural networks, analog storage element, digital potentiometers and single-transistor DACs.

History

The first report of a floating-gate MOSFET was made by Kahng and Sze, and dates back to 1967. The first application of the FGMOS was to store digital data in EEPROM, EPROM and flash memories. However, the current interest in FGMOS circuits started from developing large-scale computations in neuromorphic systems, which are inherently analog.

In 1989 Intel employed the FGMOS as an analog nonvolatile memory element in its ETANN chip, demonstrating the potential of using FGMOS devices for applications other than digital memory.

Three research accomplishments laid the groundwork for much of the current FGMOS circuit development:

- Thomsen and Brooke's demonstration and use of electron tunneling in a standard CMOS double-poly process allowed many researchers to investigate FGMOS circuits concepts without requiring access to specialized fabrication processes.
- The *ν*MOS, or neuron-MOS, circuit approach by Shibata and Ohmi provided the initial inspiration and framework to use capacitors for linear computations. These researchers concentrated on the FG circuit properties instead of the device properties, and used either UV light to equalize charge, or simulated FG elements by opening and closing MOSFET switches.
- Carver Mead›s adaptive retina gave the first example of using continuously-operating FG programming/erasing techniques, in this case UV light, as the backbone of an adaptive circuit technology.

Structure

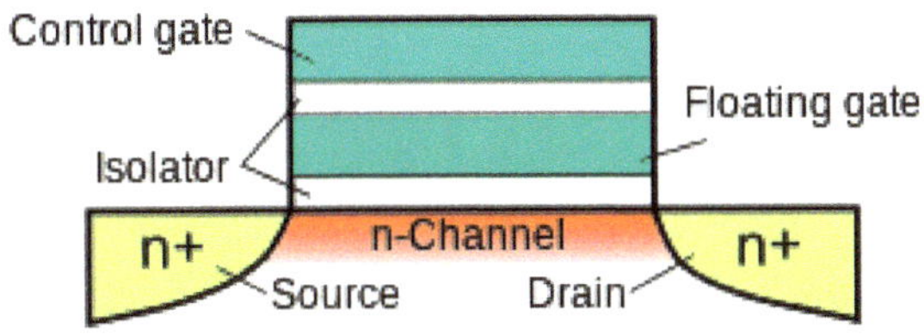

A cross-section of a floating-gate transistor

An FGMOS can be fabricated by electrically isolating the gate of a standard MOS transistor, so that there are no resistive connections to its gate. A number of secondary gates or inputs are then deposited above the floating gate (FG) and are electrically isolated from it. These inputs are only capacitively connected to the FG, since the FG is completely surrounded by highly resistive material. So, in terms of its DC operating point, the FG is a floating node.

For applications where the charge of the FG needs to be modified, a pair of small extra transistors are added to each FGMOS transistor to conduct the injection and tunneling operations. The gates of every transistor are connected together; the tunneling transistor has its source, drain and bulk terminals interconnected to create a capacitive tunneling structure. The injection transistor is connected normally and specific voltages are applied to create hot carriers that are then injected via an electric field into the floating gate.

FGMOS transistor for purely capacitive use can be fabricated on N or P versions. For charge modification applications, the tunneling transistor (and therefore the operating FGMOS) needs to be embedded into a well, hence the technology dictates the type of FGMOS that can be fabricated.

Modeling

Large Signal DC

The equations modeling the DC operation of the FGMOS can be derived from the equations that describe the operation of the MOS transistor used to build the FGMOS. If it is possible to determine the voltage at the FG of an FGMOS device, it is then possible to express its drain to source current using standard MOS transistor models. Therefore, to derive a set of equations that model the large signal operation of an FGMOS device, it is necessary to find the relationship between its effective input voltages and the voltage at its FG.

Small Signal

An N-input FGMOS device has $N-1$ more terminals than a MOS transistor, and therefore, $N+2$ small signal parameters can be defined: N effective input transconductances, an output transconductance and a bulk transconductance. Respectively:

$$g_{mi} = \frac{C_i}{C_T} g_m \quad \text{for} \quad i = [1, N]$$

$$g_{dsF} = g_{ds} + \frac{C_{GD}}{C_T} g_m$$

$$g_{mbF} = g_{mb} + \frac{C_{GB}}{C_T} g_m$$

where C_T is the total capacitance seen by the floating gate. These equations show two drawbacks of the FGMOS compared with the MOS transistor:

- Reduction of the input transconductance
- Reduction of the output resistance

Simulation

Under normal conditions, a floating node in a circuit represents an error because its initial condition is unknown unless it is somehow fixed. This generates two problems: first, it is not straightforward to simulate these circuits; and second, an unknown amount of charge might stay trapped at the floating gate during the fabrication process which will result in an unknown initial condition for the FG voltage.

Among the many solutions proposed for the computer simulation, one of the most promising methods is an Initial Transient Analysis (ITA) proposed by Rodriguez-Villegas, where the FGs are set to zero volts or a previously known voltage based on the measurement of the charge trapped in the FG after the fabrication process. A transient analysis is then run with the supply voltages set to their final values, letting the outputs evolve normally. The values of the FGs can then be extracted and used for posterior small-signal simulations, connecting a voltage supply with the initial FG value to the floating gate using a very-high-value inductor.

Applications

The usage and applications of the FGMOS can be broadly classified in two cases. If the charge in the floating gate is not modified during the circuit usage, the operation is capacitively coupled.

In the capacitively coupled regime of operation, the net charge in the floating gate is not modified. Examples of application for this regime are single transistor adders, DACs, multipliers and logic functions, variable threshold inverters,

Using the FGMOS as a programmable charge element, it is commonly used for non-volatile storage such as flash, EPROM and EEPROM memory. In this context, floating-gate MOSFETs are useful because of their ability to store an electrical charge for extended periods of time without a connection to a power supply. Other applications of the FGMOS are neuronal computational element in neural networks, analog storage element and e-pots.

BOOTSTRAPPING

In the field of electronics, a bootstrap circuit is one where part of the output of an amplifier stage is applied to the input, so as to alter the input impedance of the amplifier. When applied deliberately, the intention is usually to increase rather than decrease the impedance. Generally, any technique where part of the output of a system is used at startup is described as bootstrapping.

In the domain of MOSFET circuits, "bootstrapping" is commonly used to mean pulling up the operating point of a transistor above the power supply rail. The same term has been used somewhat more generally for dynamically altering the operating point of an operational amplifier (by shifting both its positive and negative supply rail) in order to increase its output voltage swing (relative to the ground). In the sense used in this paragraph, bootstrapping an operational amplifier means "using a signal to drive the reference point of the op-amp's power supplies". A more sophisticated use of this rail bootstrapping technique is to alter the non-linear C/V characteristic of the inputs of a JFET op-amp in order to decrease its distortion.

Input Impedance

In analog circuit designs a bootstrap circuit is an arrangement of components deliberately intended to alter the input impedance of a circuit. Usually it is intended to increase the impedance, by using a small amount of positive feedback, usually over two stages. This was often necessary in the early days of bipolar transistors, which inherently have quite a low input impedance. Because the feedback is positive, such circuits can suffer from poor stability and noise performance compared to ones that don't bootstrap.

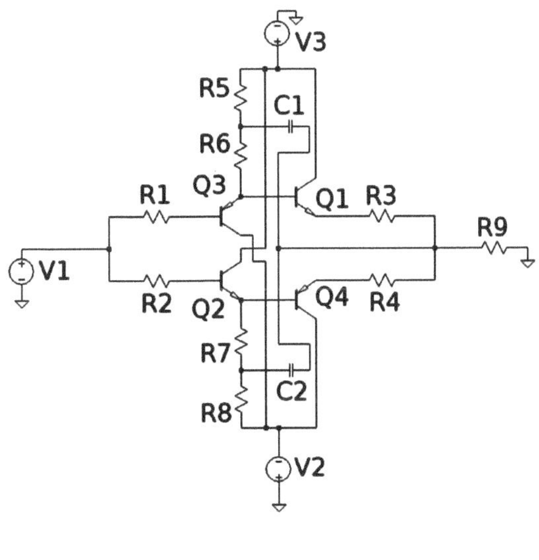

Bootstrap capacitors C1 and C2 in a BJT emitter follower circuit

Negative feedback may alternatively be used to bootstrap an input impedance, causing the apparent impedance to be reduced. This is seldom done deliberately, however, and is normally an unwanted result of a particular circuit design. A well-known example of this is the Miller effect, in which an unavoidable feedback capacitance appears increased (i.e. its impedance appears reduced) by negative feedback. One popular case where this *is* done deliberately is the Miller compensation technique for providing a low-frequency pole inside an integrated circuit. To minimize the size of the necessary capacitor, it is placed between the input and an output which swings in the opposite direction. This bootstrapping makes it act like a larger capacitor to ground.

Driving MOS Transistors

A N-MOSFET/IGBT needs a significantly positive charge ($V_{GS} > V_{\mathrm{th}}$) applied to the gate in order to turn on. Using only N-channel MOSFET/IGBT devices is a common cost reduction method due largely to die size reduction (there are other benefits as well). However, using nMOS devices in place of pMOS devices means that a voltage higher than the power rail supply (V+) is needed in order to bias the transistor into linear operation (minimal current limiting) and thus avoid significant heat loss.

A bootstrap capacitor is connected from the supply rail (V+) to the output voltage. Usually the source terminal of the N-MOSFET is connected to the cathode of a recirculation diode allowing for efficient management of stored energy in the typically inductive load. Due to the charge storage characteristics of a capacitor, the bootstrap voltage will rise above (V+) providing the needed gate drive voltage.

A MOSFET/IGBT is a voltage-controlled device which, in theory, will not have any gate current. This makes it possible to utilize the charge inside the capacitor for control purposes. However, eventually the capacitor will lose its charge due to parasitic gate current and non-ideal (i.e. finite) internal resistance, so this scheme is only used where there is a steady pulse present. This is because the pulsing action allows for the capacitor to discharge (at least partially if not completely). Most control schemes that use a bootstrap capacitor force the high side driver (N-MOSFET) off for a minimum time to allow for the capacitor to refill. This means that the duty cycle will always need to be less than 100% to accommodate for the parasitic discharge unless the leakage is accommodated for in another manner.

Switch-Mode Power Supplies

In switch-mode power supplies, the regulation circuits are powered from the output. To start the power supply, a leakage resistance can be used to trickle-charge the supply rail for the control circuit to start it oscillating. This approach is less costly and more efficient than providing a separate linear power supply just to start the regulator circuit.

Output Swing

AC amplifiers can use bootstrapping to increase output swing. A capacitor (usually referred as *bootstrap capacitor*) is connected from the output of the amplifier to the bias circuit, providing bias voltages that exceed the power supply voltage. Emitter followers can provide rail-to-rail output in this way, which is a common technique in class AB audio amplifiers.

Digital Integrated Circuits

Within an integrated circuit a bootstrap method is used to allow internal address and clock distribution lines to have an increased voltage swing. The bootstrap circuit uses a coupling capacitor, formed from the gate/source capacitance of a transistor, to drive a signal line to slightly greater than the supply voltage.

DIMMER

Dimmers are devices used to lower the brightness of a light. By changing the voltage waveform applied to the lamp, it is possible to lower the intensity of the light output. Although variable-voltage devices are used for various purposes, the term dimmer is generally reserved for those intended to control light output from resistive incandescent, halogen, and (more recently) compact fluorescent lights (CFLs) and light-emitting diodes (LEDs). More specialized equipment is needed to dim fluorescent, mercury vapor, solid state and other arc lighting.

Dimmers range in size from small units the size of a light switch used for domestic lighting to high power units used in large theatre or architectural lighting installations. Small domestic dimmers are generally directly controlled, although remote control systems (such as X10) are available. Modern professional dimmers are generally controlled by a digital control system like DMX or DALI. In newer systems, these protocols are often used in conjunction with ethernet.

In the professional lighting industry, changes in intensity are called "fades" and can be "fade up" or "fade down". Dimmers with direct manual control had a limit on the speed they could be varied at but this issue has been largely eliminated with modern digital units (although very fast changes in brightness may still be avoided for other reasons like lamp life).

Modern dimmers are built from semiconductors instead of variable resistors, because they have higher efficiency. A variable resistor would dissipate power as heat and acts as a voltage divider. Since semiconductor or solid-state dimmers switch quickly between a low resistance "on" state and a high resistance "off" state, they dissipate very little power compared with the controlled load.

A common dual dimmer manufactured by Electronic Theatre Controls(ETC).

Another dimmer by Colortran.

A residential-type dimmer switch with sliding knob to change dimness. The device is small enough to fit into a regular wall box provided for a switch.

History

Early dimmers were directly controlled through the manual manipulation of large dimmer panels. This required all power to come through the lighting control location, which could be inconvenient, inefficient and potentially dangerous for large or high-powered systems, such as those used for stage lighting.

In 1896, Granville Woods patented his "Safety Dimmer", which greatly reduced wasted energy by reducing the amount of energy generated to match desired demand rather than burning off unwanted energy.

In 1959, Joel S. Spira, who would found the Lutron Electronics Company in 1961, invented a dimmer based on a diode and a tapped autotransformer, saving energy and allowing the dimmer to be installed in a standard electrical wallbox.

In 1966, Eugene Alessio patented a light bulb socket adapter for adjusting a light level on a single light bulb using a triac. To house this device, he decided on a 2-inch round device with one end capable of being screwed into a light bulb socket and the other end able to receive a light bulb.

When solid-state dimmers came into use, analog remote control systems (such as 0-10 V lighting control systems) became feasible. The wire for the control systems was much smaller (with low current and lower danger) than the heavy power cables of previous lighting systems. Each dimmer had its own control wires, resulting in many wires leaving the lighting control location.

More recent digital control protocols such as DMX512, DALI, or one of the many Ethernet-based protocols like Art-Net, ETCnet, sACN, Pathport, ShowNet or KiNET enable the control a large number of dimmers (and other stage equipment) through a single cable.

Types of Dimmer

Rheostat Dimmer

Dimmers based on rheostats were inefficient since they would dissipate a significant portion of the power rating of the load as heat. They were large and required plenty of cooling air. Because their dimming effect depended a great deal on the total load applied to each rheostat, the load needed to be matched fairly carefully to the power rating of the rheostat. Finally, as they relied on mechanical control they were slow and it was difficult to change many channels at a time

Saltwater Dimmer

Early examples of a rheostat dimmer include a salt water dimmer or liquid rheostat; the liquid between a movable and fixed contact provided a variable resistance. The closer the contacts to each other, the more voltage was available for the light. Salt water dimmers required regular addition of water and maintenance due to corrosion; exposed parts were energized during operation, presenting a shock hazard.

Coil-Rotation Transformer

The coil-rotation transformer used a fixed-position electromagnet coil in conjunction with a variable-position coil to vary the voltage in the line by varying the alignment of the two coils. Rotated 90 degrees apart, the secondary coil is affected by two equal but opposite fields from the primary, which effectively cancel each other out and produce no voltage in the secondary.

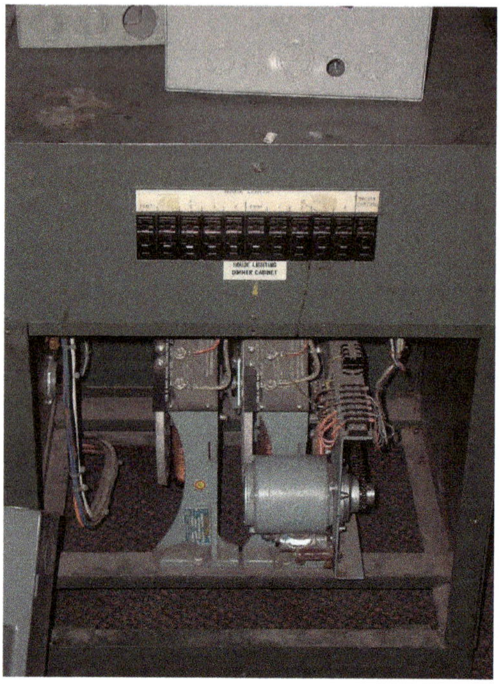

Two 6000 watt motor driven autotransformer dimmers, used for theatre auditorium lighting

These coils resembled the standard rotor and stator as used in an electric motor, except that the rotor was held against rotation using brakes and was moved to specific positions using high-torque gearing. Because the rotor did not ever turn a complete revolution, a commutator was not required and long flexible cables could be used on the rotor instead.

Autotransformer Dimmer

Variable autotransformers (trade name "Variac") were then introduced. While they were still nearly as large as rheostat dimmers, they were relatively efficient devices. Their voltage output, and so their dimming effect, was largely independent of the load applied so it was far easier to design the lighting that would be attached to each autotransformer channel. Remote control of the dimmers was still impractical, although some dimmers were equipped with motor drives that could slowly and steadily reduce or increase the brightness of the attached lamps. Autotransformers have fallen out of use for lighting but are used for other applications.

Solid-State Dimmer

Solid-state or semiconductor dimmers were introduced to solve some of these problems. Semiconductor dimmers switch on at an adjustable time (phase angle) after the start of each alternating current half-cycle, thereby altering the voltage waveform applied to lamps and so changing its RMS effective value. Because they switch instead of absorbing part of the voltage supplied, there is very little wasted power. Dimming can be almost instantaneous and is easily controlled by remote electronics. This development also made it possible to make dimmers small enough to be used in place (within the pattress) of normal domestic light switches.

A Strand CD80 thyristor dimmer rack

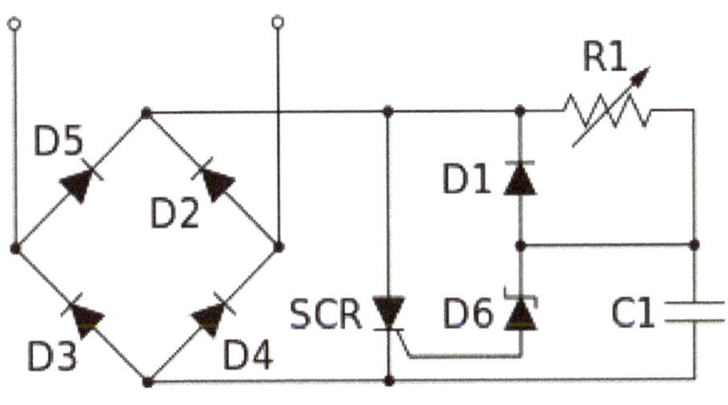

An electrical schematic for a typical SCR-based light dimmer

The switches generate some heat during switching and can cause also radio-frequency interference. Inductors or chokes are used as part of the circuitry to suppress this interference. When the dimmer is at 50% power the switches are switching their highest voltage (>325 V in Europe) and the sudden surge of power causes the coils on the inductor to move, creating buzzing sound associated with some types of dimmer; this same effect can be heard in the filaments of the incandescent lamps as “singing”. The suppression circuitry may be insufficient to prevent buzzing to be heard on sensitive audio and radio equipment that share the mains supply with the lighting loads. In this case, special considerations must be made to prevent this interference. European dimmers must comply with relevant EMC legislation requirements; this involves suppressing the emissions described above to limits described in EN55104.

In the electrical schematic to the right, a typical silicon-controlled rectifier (SCR) based light dimmer dims the light through phase angle control. This unit is wired in series with the load. Diodes (D2, D3, D4 and D5) form a bridge which generates pulsed DC. R1 and C1 form a circuit with a time constant. As the voltage increases from zero (at the start of every halfwave) C1 will charge up. When C1 is able to make Zener diode D6 conduct and inject current into the SCR, the SCR will fire. When the SCR conducts then D1 will discharge C1 via the SCR. The SCR will shut off when the current falls to zero when the supply voltage drops at the end of the half cycle, ready for the circuit to start work on the next half cycle. This is an example of a Leading Edge Dimmer.

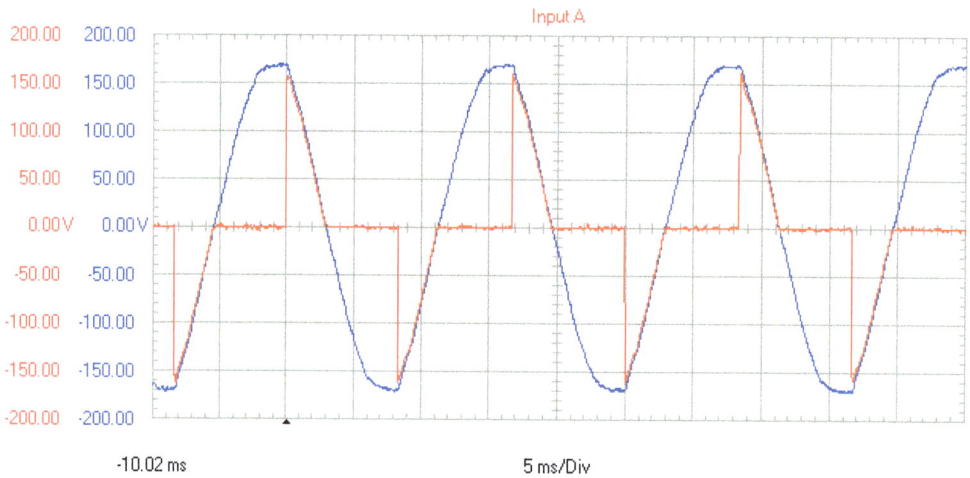

Waveform of the output voltage of a thyristor dimmer set for 60 volts RMS output, with 120 V input. The red trace shows the output device switching on about 5.5 ms after the input (blue) voltage crosses zero. Switching the thyristor on earlier in each half cycle gives a higher output voltage and brighter lights.

Dimmers based on Insulated-gate bipolar transistors (IGBTs) do away with most of the noise present in TRIACS by chopping off the falling side of the sine wave. These are termed Trailing Edge Dimmers.

An even newer, but still expensive technology is sine-wave dimming, which is implemented as high power switched-mode power supplies followed by a filter.

Control

Non domestic dimmers are usually controlled remotely by means of various protocols. Analogue dimmers usually require a separate wire for each channel of dimming carrying a voltage between 0 and 10 V. Some analogue circuitry then derives a control signal from this and the mains supply for the switches. As more channels are added to the system more wires are needed between the lighting controller and the dimmers.

A dimmer rack containing 192 dimmers, with one dimmer per circuit. The black box at the upper left is a demultiplexer.

Dimmer in a residential application with RF-based remote control

In the late 70s serial analogue protocols were developed. These multiplexed a series of analogue levels onto a single wire, with embedded clocking signal similar to a composite video signal (in the case of Strand Lighting's European D54 standard, handling 384 dimmers) or separate clocking signal (in the case of the US standard AMX192).

Digital protocols, such as DMX512 have proved to be the answer since the late 80s. In early implementations a digital signal was sent from the controller to a demultiplexer, which sat next to the dimmers. This converted the digital signal into a collection of 0 to +10 V or 0 to -10 V signals which could be connected to the individual analogue control circuits.

Modern dimmer designs use microprocessors to convert the digital signal directly into a control signal for the switches. This has many advantages, giving closer control over the dimming, and giving the opportunity for diagnostic feedback to be sent digitally back to the lighting controller.

Some dimmers in residential applications are also equipped with a radio receiver to be used as wireless light switches which can be remotely controlled by a radio transmitter.

Patching

Patching is the physical ("hard patch") or virtual ("soft patch") assignment to a circuit or channel for the purpose of control.

Hard Patch

Dimmers are usually arranged together in racks, where they can be accessed easily, and then power is run to the instruments being controlled. In architectural installations electricity is run straight from the dimmers to the lights via permanent wiring (this is called a circuit). They are hard run and cannot be changed.

A telephone-type patch bay by Hub Electric

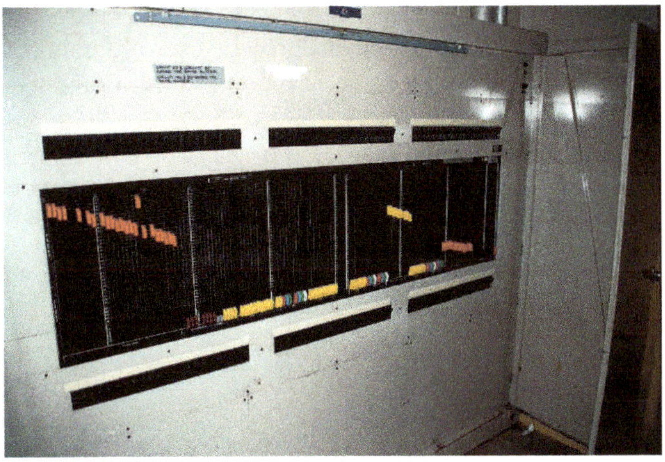

An ElectroControls Slider Type Patch Bay

However venues such as theatres demand more flexibility. To allow for changes for each show, and occasionally during shows, theatres sometimes install circuits run permanently to sockets around the theatre. Instead of these circuits going directly to the dimmer they are connected to a *patch bay*. A patch bay usually sits next to the dimmers enabling the dimmers to be connected to specific circuits via a patch cable. The patch bay may also enable many circuits to be connected to one dimmer and even series connection for low-voltage lamps. Also in some theatres individual cables are run directly from the light to dimmer. The assigned connections between the circuits (either at the patch bay or in the form of individual cables) and the dimmers is known as the mains or hard patch. Most hard patch situations A control channel on the light board is fixed to a specific dimmer as in a 1-to-1 patch. This is most common in older theatres, and on a tour where dimmers will be brought in by the touring company.

Soft Patch

Most modern fixed installations do not have patch bays, instead they have a dimmer-per-circuit and patch dimmers into channels using a computerised control consoles Soft Patch.

Dimming Curves

The design of most analogue dimmers meant that the output of the dimmer was not directly proportional to the input. Instead, as the operator brought up a fader, the dimmer would dim slowly at first, then quickly in the middle, then slowly at the top. The shape of the curve resembled that of the third quarter of a sine wave. Different dimmers produced different dimmer curves, and different applications typically demanded different responses.

Television often uses a “square law” curve, providing finer control in top part of the curve, essential to allow accurate trimming of the color temperature of lighting. Theatrical dimmers tend to use a softer “S” or linear curve. Digital dimmers can be made to have whatever curve the manufacturer desires; they may have a choice between a linear relationship and selection of different curves, so that they can be matched with older analogue dimmers. Sophisticated systems provide user-programmable or nonstandard curves, and a common use of a nonstandard curve is to turn a dimmer into a "non-dim", switching on at a user defined control level.

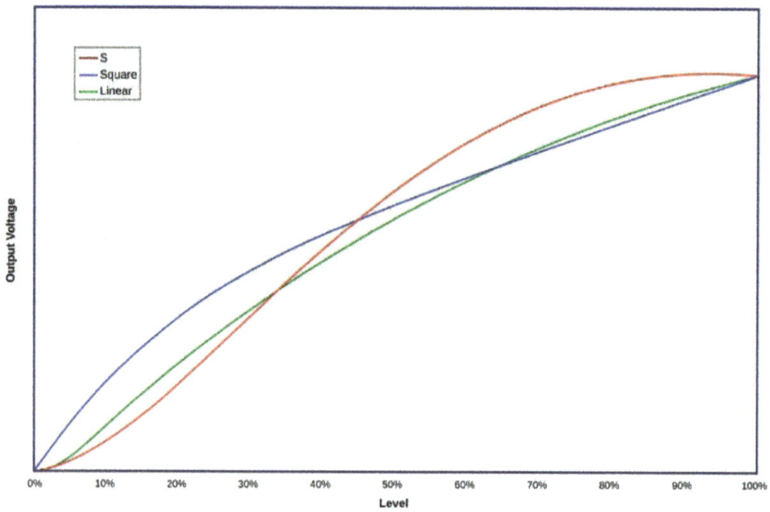

Example dimmer curves

Preheat

Switching high-intensity incandescent (filament) lamps to full power from cold can shorten their life dramatically, owing to the large inrush current that occurs. To reduce stress on the lamp filaments, dimmers may have a **preheat** function. This sets a minimum level, usually between 5% and 10%, which appears turned-off, but stops the lamp from cooling down too much. This also speeds up the lamp's reaction to sudden bursts of power that operators of rock'n'roll-style shows appreciate. The opposite of this function is sometimes called **top-set**. This limits the maximum power supplied to a lamp, which can also extend its life.

In less advanced systems, this same effect is achieved by literally pre-heating (warming) the globes before an event or performance. This is usually achieved by slowly bringing the lights up to full (or usually 90-95%) power over a period of between 1/2 to 1 hour. This is as effective as a built-in preheat function.

The Digital Revolution

Modern digital desks can emulate preheat and dimmer curves and allow a soft patch to be done in memory. This is often preferred as it means that the dimmer rack can be exchanged for another one without having to transfer complicated settings. Many different curves, or *profiles* can be programmed and used on different channels.

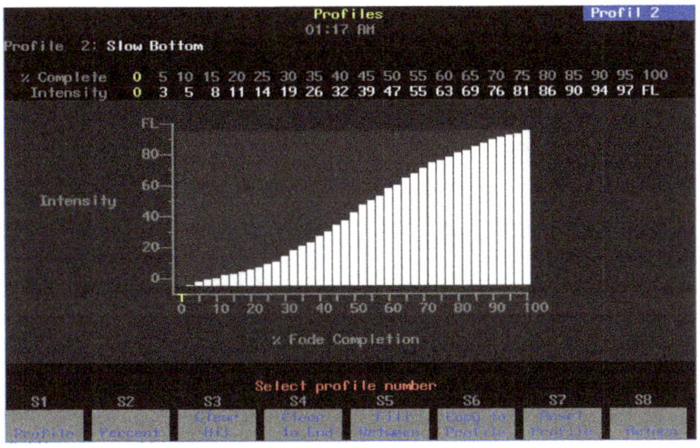

Example of an "S" curve a lightboard can soft patch

Rise Time

One measure of the quality of a leading edge dimmer is the "rise time". The rise time in this context is the amount of time it takes for the cut part of the waveform to get from zero to the instantaneous output voltage. In the waveform above it is a measure of the slope of the almost vertical edge of the red trace. Typically it is measured in tens to hundreds of microseconds. A longer rise time reduces the noise of the dimmer and the lamp as well as extending the life of the lamp. A longer rise time also reduces the electromagnetic interference produced by the dimmer. Unsurprisingly, a longer rise time is more expensive to implement than a short one, this is because the size of choke has to be increased. Newer dimming methods can help minimize such problems.

GATE TURN-OFF THYRISTOR

A gate turn-off thyristor (GTO) is a special type of thyristor, which is a high-power semiconductor device. It was invented at General Electric. GTOs, as opposed to normal thyristors, are fully controllable switches which can be turned on and off by their third lead, the GATE lead.

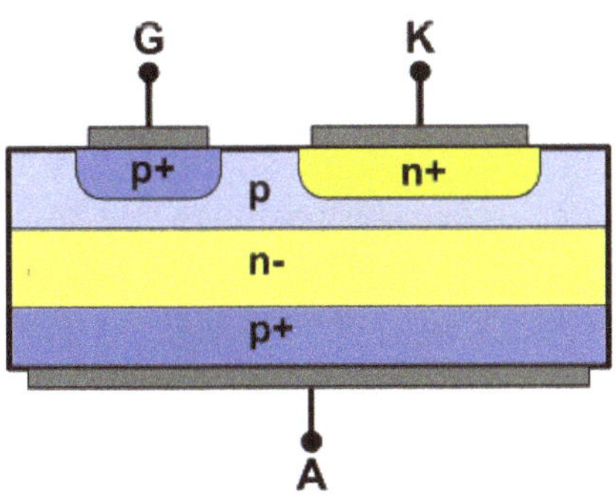

Gate turn-off thyristor GTO.

Device Description

Normal thyristors (silicon-controlled rectifiers) are not fully controllable switches (a "fully controllable switch" can be turned on and off at will). Thyristors can only be turned ON and cannot be turned OFF. Thyristors are switched ON by a gate signal, but even after the gate signal is de-asserted (removed), the thyristor remains in the ON-state until any turn-off condition occurs (which can be the application of a reverse voltage to the terminals, or when the current flowing through (forward current) falls below a certain threshold value known as the "holding current"). Thus, a thyristor behaves like a normal semiconductor diode after it is turned on or "fired".

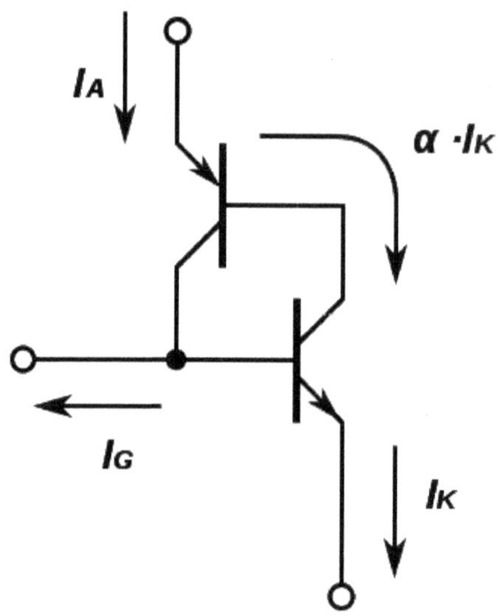

Equivalent circuit of a GTO thyristor

The GTO can be turned on by a gate signal, and can also be turned off by a gate signal of negative polarity.

Turn on is accomplished by a "positive current" pulse between the gate and cathode terminals. As the gate-cathode behaves like PN junction, there will be some relatively small voltage between the terminals. The turn on phenomenon in GTO is however, not as reliable as an SCR (thyristor) and small positive gate current must be maintained even after turn on to improve reliability.

Turn off is accomplished by a "negative voltage" pulse between the gate and cathode terminals. Some of the forward current (about one-third to one-fifth) is "stolen" and used to induce a cathode-gate voltage which in turn induces the forward current to fall and the GTO will switch off (transitioning to the 'blocking' state.)

GTO thyristors suffer from long switch off times, whereby after the forward current falls, there is a long tail time where residual current continues to flow until all remaining charge from the device is taken away. This restricts the maximum switching frequency to approx 1 kHz. It may be noted however, that the turn off time of a GTO is approximately ten times faster than that of a comparable SCR.

To assist with the turn-off process, GTO thyristors are usually constructed from a large number (hundreds or thousands) of small thyristor cells connected in parallel.

Characteristic	**Description**	**Thyristor (1600 V, 350 A)**	**GTO (1600 V, 350 A)**
$V_{T\,ON}$	On state voltage drop	1.5 V	3.4 V
t_{on}, Ig_{on}	Turn on time, gate current	8 µs,200 mA 2 µs,2 A	
t_{off}	Turn off time	150 µs	15 µs

Comparison of an SCR and GTO of same rating.

A *distributed buffer gate turn-off thyristor* (DB-GTO) is a thyristor with additional PN layers in the drift region to reshape the field profile and increase the voltage blocked in the off state. Compared to a typical PNPN structure of a conventional thyristor, this thyristor would be a PN-PN-PN type structure in here.

Reverse Bias

GTO thyristors are available with or without reverse blocking capability. Reverse blocking capability adds to the forward voltage drop because of the need to have a long, low doped P1 region.

GTO thyristors capable of blocking reverse voltage are known as Symmetrical GTO thyristors, abbreviated S-GTO. Usually, the reverse blocking voltage rating and forward blocking voltage rating are the same. The typical application for symmetrical GTO thyristors is in current source inverters.

GTO thyristors incapable of blocking reverse voltage are known as asymmetrical GTO thyristors, abbreviated A-GTO, and are generally more common than Symmetrical GTO thyristors. They typically have a reverse breakdown rating in the tens of volts. A-GTO thyristors are used where either a reverse conducting diode is applied in parallel (for example, in voltage source inverters) or where reverse voltage would never occur (for example, in switching power supplies or DC traction choppers).

GTO thyristors can be fabricated with a reverse conducting diode in the same package. These are known as RCGTO, for Reverse Conducting GTO thyristor.

Safe Operating Area

Unlike the insulated gate bipolar transistor (IGBT), the GTO thyristor requires external devices ("snubber circuits") to shape the turn on and turn off currents to prevent device destruction.

During turn on, the device has a maximum dI/dt rating limiting the rise of current. This is to allow the entire bulk of the device to reach turn on before full current is reached. If this rating is exceeded, the area of the device nearest the gate contacts will overheat and melt from over current. The rate of dI/dt is usually controlled by adding a saturable reactor (turn-on snubber), although turn-on dI/dt is a less serious constraint with GTO thyristors than it is with normal thyristors, because of the way the GTO is constructed from many small thyristor cells in parallel. Reset of the saturable reactor usually places a minimum off time requirement on GTO based circuits.

During turn off, the forward voltage of the device must be limited until the current tails off. The limit is usually around 20% of the forward blocking voltage rating. If the voltage rises too fast at turn off, not all of the device will turn off and the GTO will fail, often explosively, due to the high voltage and current focused on a small portion of the device. Substantial snubber circuits are added around the device to limit the rise of voltage at turn off. Reseting the snubber circuit usually places a minimum on time requirement on GTO based circuits.

The minimum on and off time is handled in DC motor chopper circuits by using a variable switching frequency at the lowest and highest duty cycle. This is observable in traction applications where the frequency will ramp up as the motor starts, then the frequency stays constant over most of the speed ranges, then the frequency drops back down to zero at full speed.

Applications

The main applications are in variable speed motor drives, high power inverters and traction. GTOs are increasingly being replaced by integrated gate-commutated thyristors, which are an evolutionary development of the GTO, and insulated gate bipolar transistors, which are members of the transistor family.

INTEGRATED GATE-COMMUTATED THYRISTOR

The integrated gate-commutated thyristor (IGCT) is a power semiconductor electronic device, used for switching electric current in industrial equipment. It is related to the gate turn-off (GTO) thyristor. It was jointly developed by Mitsubishi and ABB. Like the GTO thyristor, the IGCT is a fully controllable power switch, meaning that it can be turned both on and off by its control terminal (the gate). Gate drive electronics are integrated with the thyristor device.

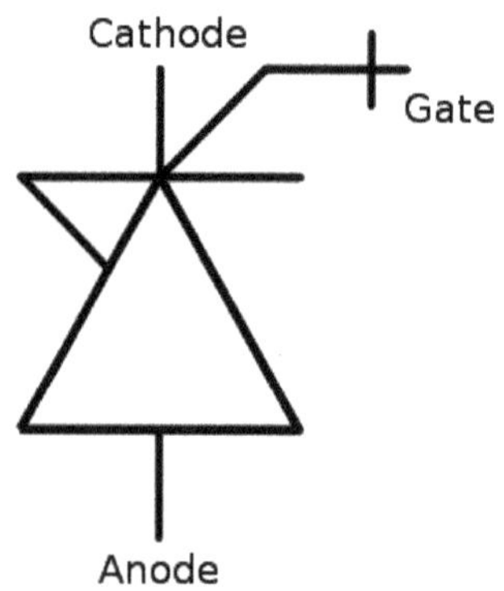

Circuit symbol for an IGCT.

Device Description

An IGCT is a special type of thyristor similar to a gate turn-off thyristor (GTO). They can be turned on and off by a gate signal, have lower conduction loss as compared to GTOs, and withstand higher rates of voltage rise (dv/dt), such that no snubber is required for most applications.

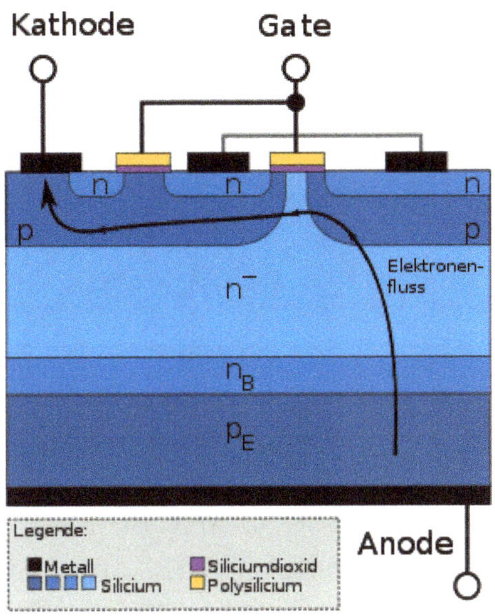

Structure of IGCT.

The structure of an IGCT is very similar to a GTO thyristor. In an IGCT, the gate turn-off current is greater than the anode current. This results in a complete elimination of minority carrier injection from the lower PN junction and faster turn-off times. The main differences are a reduction in cell size, and a much more substantial gate connection with much lower inductance in the gate drive circuit and drive circuit connection. The very high gate currents and fast dI/dt rise of the gate current mean that regular wires can not be used to connect the gate drive to the IGCT. The drive circuit PCB is integrated into the package of the device. The drive circuit surrounds the device and a large circular conductor attaching to the edge of the IGCT is used. The large contact area and short distance reduce both the inductance and resistance of the connection.

The IGCT's much faster turn-off times compared to the GTO's allows it to operate at higher frequencies—up to several kHz for very short periods of time. However, because of high switching losses, typical operating frequency is up to 500 Hz.

Reverse Bias

IGCT are available with or without reverse blocking capability. Reverse blocking capability adds to the forward voltage drop because of the need to have a long, low-doped P1 region.

IGCTs capable of blocking reverse voltage are known as symmetrical IGCT, abbreviated S-IGCT. Usually, the reverse blocking voltage rating and forward blocking voltage rating are the same. The typical application for symmetrical IGCTs is in current source inverters.

IGCTs incapable of blocking reverse voltage are known as asymmetrical IGCT, abbreviated A-IGCT. They typically have a reverse breakdown rating in the tens of volts. A-IGCTs are used where either a reverse conducting diode is applied in parallel (for example, in voltage source inverters) or where reverse voltage would never occur (for example, in switching power supplies or DC traction choppers).

Asymmetrical IGCTs can be fabricated with a reverse conducting diode in the same package. These are known as RC-IGCT, for reverse conducting IGCT.

Applications

The main applications are in variable-frequency inverters, drives and traction.

REFERENCES

1. Baliga, B. Jayant. Power Semiconductor Devices. Boston: PWS publishing Company. ISBN 0-534-94098-6.
2. Jain, Alok. Power Electronics and Its Applications. Mumbai: Penram International Publishing. ISBN 81-87972-22-X.
3. Semikron: Application Manual IGBT and MOSFET Power Modules, 2. Edition, 2015,ISLE Verlag, ISBN 978-3-938843-83-3 PDF-Version
4. Arendt Wintrich; Ulrich Nicolai; Werner Tursky; Tobias Reimann (2010) (in German), Applikationshandbuch 2015 (2. ed.), ISLE Verlag, ISBN 978-3-938843-83-3
5. Arendt Wintrich; Ulrich Nicolai; Werner Tursky; Tobias Reimann (2015). Application Manual 2015 (PDF-Version) (2. ed.). ISLE Verlag. ISBN 978-3-938843-83-3.
6. Paul Horowitz and Winfield Hill (1989). The Art of Electronics (2nd ed.). Cambridge University Press. ISBN 978-0-521-37095-0.
7. Juin Jei Liou and Jiann S. Yuan (1998). Semiconductor Device Physics and Simulation. Springer. ISBN 0-306-45724-5.
8. General Electric (1962). Transistor Manual (6th ed.). p. 12. "If the principle of space charge neutrality is used in the analysis of the transistor, it is evident that the collector current is controlled by means of the positive charge (hole concentration) in the base region. ... When a transistor is used at higher frequencies, the fundamental limitation is the time it takes the carriers to diffuse across the base region..." (same in 4th and 5th editions)
9. Paolo Antognetti and Giuseppe Massobrio (1993). Semiconductor Device Modeling with Spice. McGraw–Hill Professional. ISBN 0-07-134955-3.
10. Alphonse J. Sistino (1996). Essentials of electronic circuitry. CRC Press. p. 64.

ISBN 978-0-8247-9693-8.

11. Alphonse J. Sistino (1996). Essentials of electronic circuitry. CRC Press. p. 102. ISBN 978-0-8247-9693-8.
12. D.V. Morgan, Robin H. Williams (Editors) (1991). Physics and Technology of Heterojunction Devices. London: Institution of Electrical Engineers (Peter Peregrinus Ltd.). ISBN 0-86341-204-1.
13. Peter Ashburn (2003). SiGe Heterojunction Bipolar Transistors. New York: Wiley. Chapter 10. ISBN 0-470-84838-3.
14. Paul Horowitz and Winfield Hill (1989). The Art of Electronics (2nd ed.). Cambridge University Press. pp. 62–66. ISBN 978-0-521-37095-0.
15. "1947: Invention of the Point-Contact Transistor - The Silicon Engine - Computer History Museum". Retrieved August 10, 2016.
16. "1948: Conception of the Junction Transistor - The Silicon Engine - Computer History Museum". Retrieved August 10, 2016.
17. Third case study – the solid state advent Archived September 27, 2007, at the Wayback Machine. (PDF)
18. Transistor Museum, Historic Transistor Photo Gallery, Bell Labs Type M1752
19. Morris, Peter Robin (1990). "4.2". A History of the World Semiconductor Industry. IEE History of Technology Series 12. London: Peter Peregrinus Ltd. p. 29. ISBN 0-86341-227-0.
20. "Transistor Museum Photo Gallery RCA TA153". Retrieved August 10, 2016.
21. High Speed Switching Transistor Handbook (2nd ed.). Motorola. 1963. p. 17.
22. Transistor Museum, Historic Transistor Photo Gallery, Western Electric 3N22
23. The Tetrode Power Transistor PDF
24. "Transistor Museum Photo Gallery Philco A01 Germanium Surface Barrier Transistor". Retrieved August 10, 2016.
25. "Transistor Museum Photo Gallery Germanium Surface Barrier Transistor". Retrieved August 10, 2016.
26. Herb's Bipolar Transistors IEEE Transactions on Electron Devices, vol. 48, no. 11, November 2001 PDF
27. Influence of Mobility and Lifetime Variations on Drift-Field Effects in Silicon-Junction Devices PDF
28. "Transistor Museum Photo Gallery Bell Labs Prototype Diffused Base Germanium Silicon Transistor". Retrieved August 10, 2016.
29. "Transistor Museum Photo Gallery Fairchild 2N1613 Early Silicon Planar Transistor". Retrieved August 10, 2016.
30. "1960: Epitaxial Deposition Process Enhances Transistor Performance - The Silicon Engine - Computer History Museum". Retrieved August 10, 2016.
31. J.J. Ebers and J.L Moll (1954) "Large-signal behavior of junction transistors", Proceedings of the Institute of Radio Engineers, 42 (12) : 1761–1772.

32. Adel S. Sedra and Kenneth C. Smith (1987). Microelectronic Circuits, second ed. p. 903. ISBN 0-03-007328-6.
33. A.S. Sedra and K.C. Smith (2004). Microelectronic Circuits (5th ed.). New York: Oxford. Eqs. 4.103–4.110, p. 305. ISBN 0-19-514251-9.
34. H. K. Gummel and R. C. Poon, "An integral charge control model of bipolar transistors", Bell Syst. Tech. J., vol. 49, pp. 827–852, May–June 1970
35. "Bipolar Junction Transistors". Retrieved August 10, 2016.
36. A.S. Sedra and K.C. Smith (2004). Microelectronic Circuits (5th ed.). New York: Oxford. p. 509. ISBN 0-19-514251-9.
37. http://www.silvaco.com/content/kbase/smartspice_device_models.pdf
38. Gennady Gildenblat, ed. (2010). Compact Modeling: Principles, Techniques and Applications. Springer Science & Business Media. Part II: Compact Models of Bipolar Junction Transistors, pp. 167-267 cover Mextram and HiCuM in-depth. ISBN 978-90-481-8614-3.
39. Michael Schröter (2010). Compact Hierarchical Bipolar Transistor Modeling with Hicum. World Scientific. ISBN 978-981-4273-21-3.
40. http://joerg-berkner.de/Fachartikel/pdf/2002 ICCAP UM Berkner Compact_Models_4_BJTs.pdf
41. "IC Temperature Sensors Find the Hot Spots - Application Note - Maxim". maxim-ic.com. February 21, 2002. Retrieved August 10, 2016.
42. IEEE Standard 100 Authoritative Dictionary of IEEE Standards Terms, Seventh Edition, IEEE Press, 2000 ISBN 0-7381-2601-2 page 123
43. John P. Uyemura (1999). CMOS Logic Circuit Design. Springer Science & Business Media. p. 319. ISBN 978-0-7923-8452-6.
44. Marcel J.M. Pelgrom (2012). Analog-to-Digital Conversion (2nd ed.). Springer Science & Business Media. pp. 210–211. ISBN 978-1-4614-1371-4.
45. Grayson King and Tim Watkins (May 13, 1999). "Bootstrapping your op amp yields wide voltage swings" (PDF). EDN: 117–129.
46. Jerald Graeme (1994). "Op-amp distortion measurement bypasses test-equipment limitations". In Ian Hickman and Bill Travis. The EDN Designer's Companion. Butterworth-Heinemann. p. 205. ISBN 978-0-7506-1721-5.
47. Walt Jung. "Bootstrapped IC Substrate Lowers Distortion in JFET Op Amps" (PDF). Analog Devices application note AN-232.
48. Douglas Self (2014). Small Signal Audio Design (2nd ed.). Focal Press. pp. 136–142. ISBN 978-1-134-63513-9.
49. Raymond A. Mack Demystifying switching power supplies Newnes, 2005 ISBN 0-7506-7445-8, page 121
50. William J. Dally, John W. Poulton, Digital systems engineering, Cambridge University Press, 1998 ISBN 0-521-59292-5 pages 190-191

CHAPTER 3

DC to DC Choppers

Chapter Contents

- **Introduction**
- **Classification**
- **Applications**
- **Control Strategies**
- **Chopper Amplifiers**
- **Braking Chopper**
- **Vibrator**
- **Multivibrator**

A chopper is a high-speed on/off semiconductor switch. It connects source to load and disconnects the load from source at high-speed. In other words, the principle of chopper is application of fixed dc voltage intermittently to the load. This is achieved by continuously triggering ON and triggering OFF the power switch(SCR) at rapid speed. The duration for which the SCR remains in ON and OFF states are called ON time and OFF time respectively. By varying the ON time and OFF time of the SCR, the average voltage across the load can be varied.

The terms DC–DC converters and choppers are one and same. In the texts usually these terms are interchanged. The Choppers can be operated in either a continuous or discontinuous current conduction mode. They can be built with and without electrical isolation.

INTRODUCTION

In electronics, a chopper circuit is used to refer to numerous types of electronic switching devices and circuits used in power control and signal applications. A chopper is a switching device that converts fixed DC input to a variable DC output voltage directly. Essentially, a chopper is an electronic switch that is used to interrupt one signal under the control of another.

In power electronics applications, since the switching element is either fully on or fully off, its losses are low, and the circuit can provide high efficiency. However, the current supplied to the load is discontinuous and may require smoothing or a high switching frequency to avoid undesirable effects. In signal processing circuits, use of a chopper stabilizes a system against drift of electronic components; the original signal can be recovered after amplification or other processing by a synchronous demodulator that essentially un-does the "chopping" process.

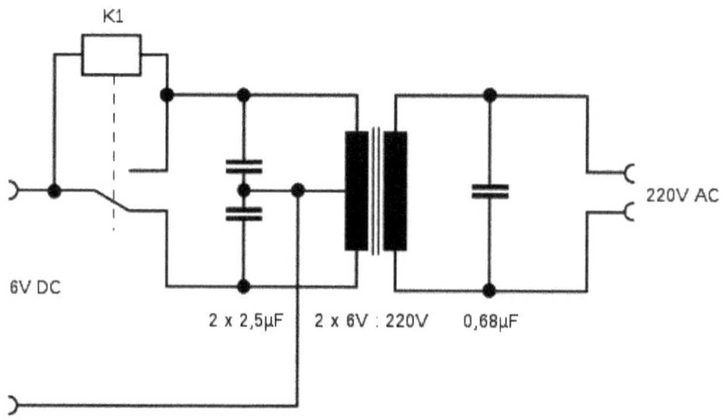

Schematic of an inverter using a vibrator as a chopper.

CLASSIFICATION

Choppers may be classified on several bases.

- On basis of input and output voltage levels:
  - Step-down chopper
    - class A
    - class B
    - class C(combination of A&B)
    - class D
    - class E
  - Step-up chopper
    - class B *explanation of class A, C,D,E chopper
  - Comparison between step up and step down chopper:

Sr no	parameters	Step down chopper	Step up chopper
1	Range of output voltage	0 to V volts	V to +∞ volts
2	Position of chopper switch	In series with load	In parallel with load
3	Expression for output voltage	VL dc = D x V volts	Vo = V/ (1 – D) volts
4	External inductance	Not required	Required for boosting the output voltage
5	Use	For motoring operation, for motor load	For regenerative braking for motor load.
6	Type of chopper	Single quadrant	Single quadrant
7	Quadrant of operation	1st quadrant	1st quadrant
8	Applications	Motor speed control	Battery charging/voltage boosters

On basis of circuit operation:

- First quadrant
- Two quadrant
- Four quadrant

■ On basis of commutation method:

- Voltage commutated
- Current commutated
- Load commutated
- Impulse commutated

APPLICATIONS

Most modern uses also use alternative nomenclature which helps to clarify which particular type of circuit is being discussed. These include:

- switched mode power supplies, including DC to DC converters.
- Speed controllers for DC motors
- Class D Electronic amplifiers
- Switchcd capacitor filtcrs
- Variable-frequency drives
- D.C. motor speed control
- D.C. voltage boosting
- Battery-operated electric cars
- Battery-operated appliances
- Battery chargers

CONTROL STRATEGIES

For all the chopper configurations operating from a fixed DC input voltage, the average value of the output voltage is controlled by periodic opening and closing of the switches used in the chopper circuit. The average output voltage can be controlled by different techniques namely:

- Pulse-width modulation
- Frequency modulation
- Variable frequency, variable pulse width
- CLC control

In pulse-width modulation the switches are turned on at a constant chopping frequency. The total time period of one cycle of output waveform is constant. The average output voltage is directly proportional to the ON time of chopper. The ratio of ON time to total time is defined as duty cycle. It can be varied between 0 and 1 or between 0 and 100%. Pulse-width modulation (PWM), or pulse-duration modulation (PDM), is a technique used to encode a message into a pulsing signal. Although this modulation technique can be used to encode information for transmission, its main use is to allow the control of the power supplied to electrical devices, especially to inertial loads such as motors. The average value of voltage (and current) fed to the load is controlled by turning the switch between supply and load on and off at a fast rate. The longer the switch is on compared to the off periods, the higher the total power supplied to the load. The PWM switching frequency has to be much higher than what would affect the load (the device that uses the power), which is to say that the resultant waveform perceived by the load must be as smooth as possible. Typically switching has to be done several times a minute in an electric stove, 120 Hz in a lamp dimmer, from few kilohertz (kHz) to tens of kHz for a motor drive and well into the tens or hundreds of kHz in audio amplifiers and computer power supplies.

In frequency modulation, pulses of a fixed amplitude and duration are generated and the average value of output is adjusted by changing how often the pulses are generated.

Variable pulse width and frequency combines both changes in the pulse width and repetition rate.

CHOPPER AMPLIFIERS

One classic use for a chopper circuit and where the term is still in use is in chopper amplifiers. These are DC amplifiers. Some types of signals that need amplifying can be so small that an incredibly high gain is required, but very high gain DC amplifiers are much harder to build with low offset and 1/f noise, and reasonable stability and bandwidth. It's much easier to build an AC amplifier instead. A chopper circuit is used to break up the input signal so that it can be processed as if it were an AC signal, then integrated back to a DC signal at the output. In this way, extremely small DC signals can be amplified. This approach is often used in electronic instrumentation where stability and accuracy are essential; for example, it is possible using these techniques to construct pico-voltmeters and Hall sensors.

The input offset voltage of amplifiers becomes important when trying to amplify small signals with very high gain. Because this technique creates a very low input offset voltage amplifier, and because this input offset voltage does not change much with time and temperature, these techniques are also called "zero-drift" amplifiers (because there is no drift in input offset voltage with time and temperature). Related techniques that also give these zero-drift advantages are auto-zero and chopper-stabilized amplifiers.

Auto-zero amplifiers use a secondary auxiliary amplifier to correct the input offset voltage of a main amplifier. Chopper-stabilized amplifiers use a combination of auto-zero and chopper techniques to give some excellent DC precision specifications.

Some example chopper and auto-zero amplifiers are LTC2050, MAX4238/MAX4239 and OPA333.

Examples and Circuits

The Keithley Instruments Model Nanovolt Null Detector and the 147 Leeds & Northrup 9838-1 Guarded Nanovolt Detector both use mechanical choppers. The choppers are like buzzers and you can hear them at an audio frequency.

The Keithley manual is available at their website. And the Leeds & Northrup has a similar circuit.

The Keithley has been used to look for the I-R DC voltage of a superconductor. The system noise was such that a voltage as low as 10 nanovolts could be detected, and none was found.

A Leeds & Northrup was bench tested and found to be capable of detecting 50 nanovolts.

BRAKING CHOPPER

Braking choppers sometimes also referred to as braking unit are used in the DC voltage intermediate circuits of frequency converters to control voltage when the load feeds energy back to the intermediate circuit. This arises, for example, when a magnetized motor is being rotated by an overhauling load and so functions as a generator feeding power to the DC voltage intermediate circuit. They are an application of the chopper principle, using on-off control of a switching device.

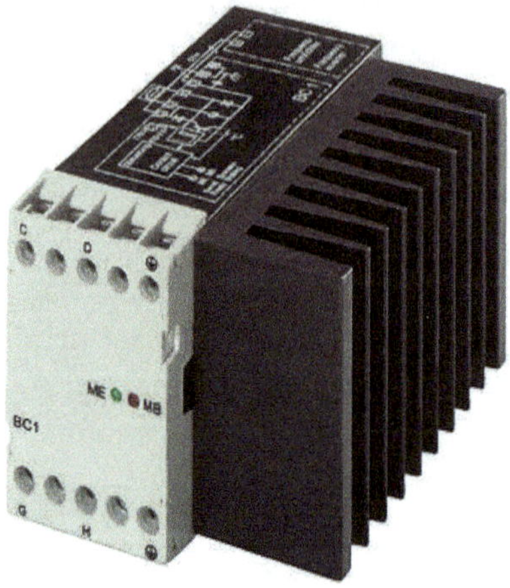

Braking chopper.

Operation

Large braking chopper installation

A braking chopper is an electrical switch that limits the DC bus voltage by switching the braking energy to a resistor where the braking energy is converted to heat. Braking choppers are automatically activated when the actual DC bus voltage exceeds a specified level depending on the nominal voltage of the variable-frequency drive.

Benefits

- Simple electrical construction and well-known technology.
- Low fundamental investment for chopper and resistor.
- The chopper works even if AC supply is lost. Braking during main power loss may be required. E.g. in elevator or other safety related applications.

Drawbacks

- The braking energy is wasted if the heated air cannot be used.
- The braking chopper and resistors require additional space.
- May require extra investments in the cooling and heat recovery system.
- Braking choppers are typically dimensioned for a certain cycle, e.g. 100% power 1/10 minutes, long braking times require more accurate dimensioning of the braking chopper.
- Increased risk of fire due to hot resistor and possible dust and chemical components in the ambient air space.
- The increased DC bus voltage level during braking causes additional voltage stress on motor insulation.

Applications

Braking choppers are inappropriate when:

- The braking cycle is needed only occasionally.
- The amount of braking energy with respect to motoring energy is extremely small.
- The ambient air includes substantial amounts of dust or other potentially combustible or explosive or metallic components.

Braking choppers are appropriate when:

- The braking is continuous or regularly repeated.
- The total amount of braking energy is high in respect to the motoring energy needed.
- The instantaneous braking power is high, e.g. several hundred kW for several minutes.
- Braking operation is needed during main power loss.

Flux Braking

Flux braking is another method, based on motor losses, for handling an overrunning load. When braking in the drive system is needed, the motor flux and thus also the magnetizing current component used in the motor are increased. The control of flux can be easily achieved through the direct torque control principle. With DTC the inverter is directly controlled to achieve the desired torque and flux for the motor. During flux braking the motor is under DTC control which guarantees that braking can be made according to the specified speed ramp. This is very different to the DC injection braking typically used in drives. In the DC injection method DC current is injected to the motor so that control of the motor flux is lost during braking. The flux braking method based on DTC enables the motor to shift quickly from braking to motoring power when requested.

In flux braking the increased current means increased losses inside the motor. The braking power is therefore also increased although the braking power delivered to the frequency converter is not increased. The increased current generates increased losses in motor resistances. The higher the resistance value the higher the braking energy dissipation inside the motor. Typically, in low power motors(below 5 kW) the resistance value of the motor is relatively large in respect to the nominal current of the motor. The higher the power or the voltage of the motor the less the resistance value of the motor in respect to motor current. In other words, flux braking is most effective in a low power motor.

VIBRATOR

In electronics before the development of switch-mode power supplies and the introduction of semiconductor devices operating off low voltage, there was a requirement to generate voltages of about 50 to 250V DC from vehicle batteries. Electromechanical components known as vibrators were used in a circuit similar to modern solid state inverter circuits to provide a pulsating DC which could be converted to a higher voltage with a transformer, rectified, and filtered to create higher-voltage DC. This “vibrator” is essentially a relay using normally closed contacts to supply power to the relay coil, thus immediately breaking the connection, only to be reconnected very quickly through the normally closed contacts. It happens so rapidly it vibrates, and sounds like a buzzer. This same rapidly pulsing contact applies the rising and falling DC voltage to the transformer which can step it up to a higher voltage.

The primary use for this type of circuit was to operate vacuum tube radios in vehicles, but it also saw use with other mobile electronic devices with a 6 or 12V accumulator, especially in places with no mains electricity supply such as farms. These vibrator power supplies became popular in the 1940s, replacing more bulky motor-generator systems for the generation of AC voltages for such applications. Vacuum tubes require plate voltages ranging from about 45 volts to 250 volts in electronic devices such as radios. For portable radios, hearing aids and similar equipment, *B batteries* were manufactured with various voltage ratings. In order to provide the necessary voltage for a radio from the typical 6 or 12 volt DC supply available in a car or from a farm lighting battery, it was necessary to convert the steady DC supply to a pulsating DC and use a transformer to increase the voltage.

A pair of Heathkit-brand vibrators manufactured by James Electronics, with octal bases. The one on the right has been stripped of the aluminum cap so the inner components can be seen.

An electro-mechanical vibrator from the Grass Instrument Co. Used as part of a chopper amplifier in polygraph input amplifier.

Vibrators often experienced mechanical malfunctions, being constantly in motion, such as the springs losing tension, and the contact points wearing down. As tubes began to be replaced by transistor based electrical systems, the need to generate such high voltages began to diminish. Mechanical vibrators fell out of production near the end of the 20th century, but solid-state electronic vibrators are still manufactured to be backwards compatible with older units.

USE

The vibrator was a device with switch contacts on flexible metal strips. When the vibrator was operated the strips vibrated rapidly in such a way as to make contacts presenting a 6 or 12V DC from the battery in alternating sense, rapidly exchanging positive and negative, effectively generating a square wave. Unlike DC, an alternating voltage applied to the terminals of the primary winding of a transformer will induce a different (and often higher) voltage of any desired value at the secondary; this voltage can be rectified by a thermionic diode or copper-oxide or selenium rectifier, or by an additional set of mechanical contacts making the vibrator output a synchronous rectifier. The rectified output is then filtered, ultimately converting a DC voltage to a higher one, with some losses dissipated as heat.

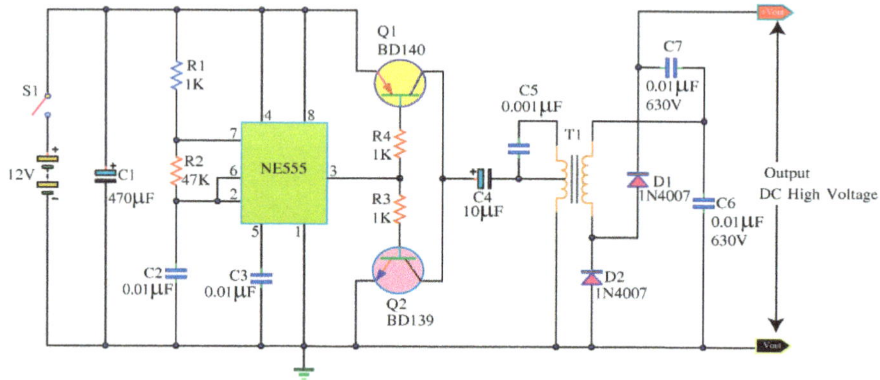

Schematic diagram of a typical circuit to convert low voltage DC to high voltage DC

This arrangement is an electromechanical inverter circuit. The vibrator switches used for this purpose wore out after a time, so they were usually encased in a steel or aluminum can with a plug for easy replacement when mounted in a socket similar to those used for tubes.

MULTIVIBRATOR

A multivibrator is an electronic circuit used to implement a variety of simple two-state devices such as relaxation oscillators, timers and flip-flops. It consists of two amplifying devices (transistors, vacuum tubes or other devices) cross-coupled by resistors or capacitors. The first multivibrator circuit, the astable multivibrator oscillator, was invented by Henri Abraham and Eugene Bloch during World War I. They called their circuit a “multivibrator” because its output waveform was rich in harmonics.

The three types of multivibrator circuits are:

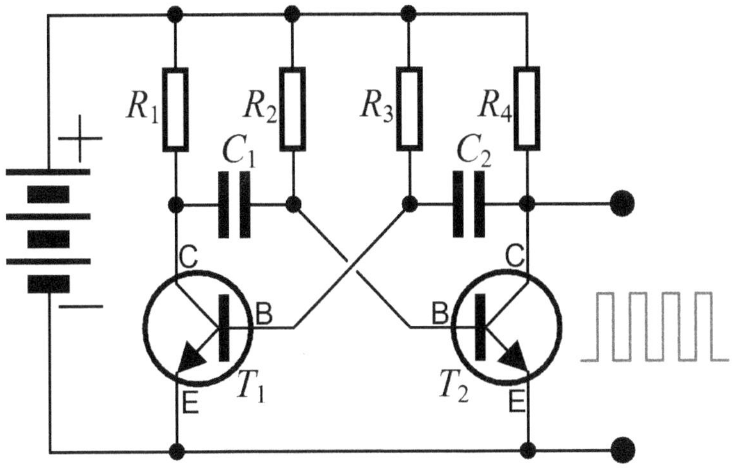

Original vacuum tube Abraham-Bloch multivibrator oscillator, from their 1919 paper

- Astable multivibrator, in which the circuit is not stable in either state —it continually switches from one state to the other. It functions as a relaxation oscillator.
- Monostable multivibrator, in which one of the states is stable, but the other state is unstable (transient). A trigger pulse causes the circuit to enter the unstable state. After entering the unstable state, the circuit will return to the stable state after a set time. Such a circuit is useful for creating a timing period of fixed duration in response to some external event. This circuit is also known as a one shot.
- Bistable multivibrator, in which the circuit is stable in either state. It can be flipped from one state to the other by an external trigger pulse. This circuit is also known as a flip-flop. It can store one bit of information, and is widely used in digital logic and computer memory.

Multivibrators find applications in a variety of systems where square waves or timed intervals are required. For example, before the advent of low-cost integrated circuits, chains of multivibrators found use as frequency dividers. A free-running multivibrator with a frequency of one-half to one-tenth of the reference frequency would accurately lock to the reference frequency. This technique was used in early electronic organs, to keep notes of different octaves accurately in tune. Other applications included early television systems, where the various line and frame frequencies were kept synchronized by pulses included in the video signal.

History

The first multivibrator circuit, the classic astable multivibrator oscillator (also called a *plate-coupled multivibrator*) was first described by Henri Abraham and Eugene Bloch in *Publication 27* of the French *Ministère de la Guerre*, and in *Annales de Physique 12, 252 (1919)*. Since it produced a square wave, in contrast to the sine wave generated by most other oscillator circuits of the time, its output contained many harmonics above the fundamental frequency, which could be used for calibrating high frequency radio circuits. For this reason Abraham and Bloch called it a *multivibrateur*. It is a predecessor of the Eccles-Jordan triggerwhich was derived from the circuit a year later.

A vacuum tube Abraham-Bloch multivibrator oscillator, France, 1920 *(small box, left)*. Its harmonics are being used to calibrate a wavemeter *(center)*.

Historically, the terminology of multivibrators has been somewhat variable:

- 1942 – multivibrator implies astable: "The multivibrator circuit (Fig. 7-6) is somewhat similar to the flip-flop circuit, but the coupling from the anode of one valve to the grid of the other is by a condenser only, so that the coupling is not maintained in the steady state."
- 1942 – multivibrator as a particular flip-flop circuit: "Such circuits were known as 'trigger' or 'flip-flop' circuits and were of very great importance. The earliest and best known of these circuits was the multivibrator."
- 1943 – flip-flop as one-shot pulse generator: "It should be noted that an essential difference between the two-valve flip-flop and the multivibrator is that the flip-flop has one of the valves biased to cutoff."
- 1949 – monostable as flip-flop: "Monostable multivibrators have also been called 'flip-flops'."
- 1949 – monostable as flip-flop: "... a flip-flop is a monostable multivibrator and the ordinary multivibrator is an astable multivibrator."

Astable Multivibrator

An astable multivibrator consists of two amplifying stages connected in a positive feedback loop by two capacitive-resistive coupling networks. The amplifying elements may be junction or field-effect transistors, vacuum tubes, operational amplifiers, or other types of amplifier. The example diagram on the right shows bipolar junction transistors.

The circuit is usually drawn in a symmetric form as a cross-coupled pair. Two output terminals can be defined at the active devices, which will have complementary states; one will have high voltage while the other has low voltage, (except during the brief transitions from one state to the other).

Operation

The circuit has two astable (unstable) states that change alternatively with maximum transition rate because of the "accelerating" positive feedback. It is implemented by the coupling capacitors that instantly transfer voltage changes because the voltage across a capacitor cannot suddenly change. In each state, one transistor is switched on and the other is switched off. Accordingly, one fully charged capacitor discharges (reverse charges) slowly thus converting the time into an exponentially changing voltage. At the same time, the other empty capacitor quickly charges thus restoring its charge (the first capacitor acts as a time-setting capacitor and the second prepares to play this role in the next state). The circuit operation is based on the fact that the forward-biased base-emitter junction of the switched-on bipolar transistor can provide a path for the capacitor restoration.

State 1 (Q1 is switched on, Q2 is switched off)

In the beginning, the capacitor C1 is fully charged (in the previous State 2) to the power supply voltage V with the polarity shown in Figure 1. Q1 is *on* and connects the left-hand positive plate of C1 to ground. As its right-hand negative plate is connected to Q2 base, a maximum negative voltage ($-V$) is applied to Q2 base that keeps Q2 firmly *off*. C1 begins discharging (reverse charging) via the high-value base resistor R2, so that the voltage of its right-hand plate (and at the base of Q2) is rising from below ground ($-V$) toward $+V$. As Q2 base-emitter junction is reverse-biased, it does not conduct, so all the current from R2 goes into C1. Simultaneously, C2 that is fully discharged and even slightly charged to 0.6 V (in the previous State 2) quickly charges via the low-value collector resistor R4 and Q1 forward-biased base-emitter junction (because R4 is less than R2, C2 charges faster than C1). Thus C2 restores its charge and prepares for the next State C2 when it will act as a time-setting capacitor. Q1 is firmly saturated in the beginning by the "forcing" C2 charging current added to R3 current. In the end, only R3 provides the needed input base current. The resistance R3 is chosen small enough to keep Q1 (not deeply) saturated after C2 is fully charged.

When the voltage of C1 right-hand plate (Q2 base voltage) becomes positive and reaches 0.6 V, Q2 base-emitter junction begins diverting a part of R2 charging current. Q2 begins conducting and this starts the avalanche-like positive feedback process as follows. Q2 collector voltage begins falling; this change transfers through the fully charged C2 to Q1 base and Q1 begins cutting off. Its collector voltage begins rising; this change transfers back through the almost empty C1 to Q2 base and makes Q2 conduct more thus sustaining the initial input impact on Q2 base. Thus the initial input change circulates along the feedback loop and grows in an avalanche-like manner until finally Q1 switches off and Q2 switches on. The forward-biased Q2 base-emitter junction fixes the voltage of C1 right-hand plate at 0.6 V and does not allow it to continue rising toward $+V$.

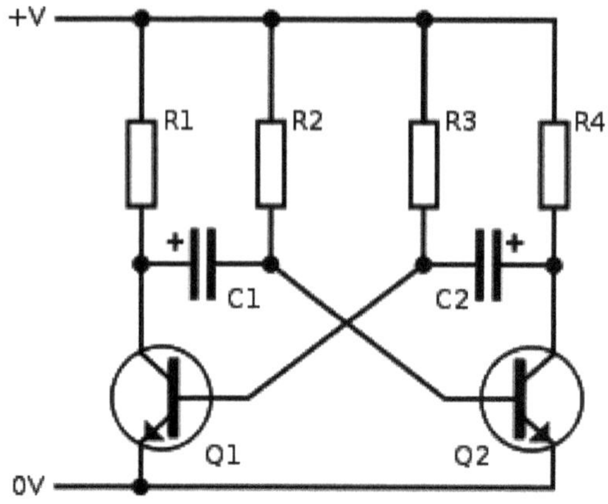

Basic BJT astable multivibrator

State 2 (Q1 is switched off, Q2 is switched on)

Now, the capacitor C2 is fully charged (in the previous State 1) to the power supply voltage V with the polarity shown in Figure 1. Q2 is *on* and connects the right-hand positive plate of C2 to ground. As its left-hand negative plate is connected to Q1 base, a maximum negative voltage ($-V$) is applied to Q1 base that keeps Q1 firmly *off*. C2 begins discharging (reverse charging) via the high-value base resistor R3, so that the voltage of its left-hand plate (and at the base of Q1) is rising from below ground ($-V$) toward $+V$. Simultaneously, C1 that is fully discharged and even slightly charged to 0.6 V (in the previous State 1) quickly charges via the low-value collector resistor R1 and Q2 forward-biased base-emitter junction (because R1 is less than R3, C1 charges faster than C2). Thus C1 restores its charge and prepares for the next State 1 when it will act again as a time-setting capacitor...and so on... (the next explanations are a mirror copy of the second part of State 1).

Monostable

In the monostable multivibrator, the one resistive-capacitive network (C_2-R_3 in figure 1) is replaced by a resistive network (just a resistor). The circuit can be thought as a 1/2 astable multivibrator. Q2 collector voltage is the output of the circuit (in contrast to the astable circuit, it has a perfect square waveform since the output is not loaded by the capacitor).

When triggered by an input pulse, a monostable multivibrator will switch to its unstable position for a period of time, and then return to its stable state. The time period monostable multivibrator remains in unstable state is given by $t = \ln(2)R_2C_1$. If repeated application of the input pulse maintains the circuit in the unstable state, it is called a *retriggerable* monostable. If further trigger pulses do not affect the period, the circuit is a *non-retriggerable* multivibrator.

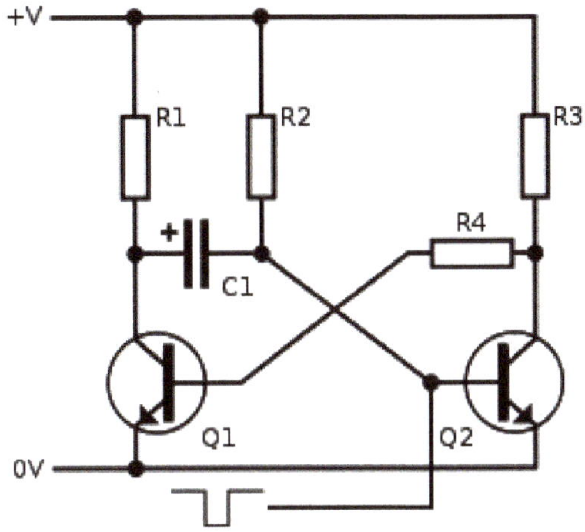

Basic BJT monostable multivibrator

For the circuit in Figure 2, in the stable state Q1 is turned off and Q2 is turned on. It is triggered by zero or negative input signal applied to Q2 base (with the same success it can be triggered by applying a positive input signal through a resistor to Q1 base). As a result, the circuit goes in State 1 described above. After elapsing the time, it returns to its stable initial state.

Bistable

In the bistable multivibrator, both the resistive-capacitive network are replaced by resistive networks (just resistors or direct coupling).

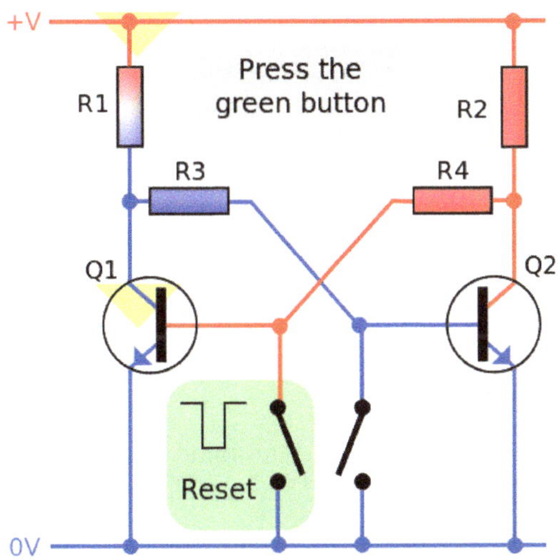

Basic animated interactive BJT bistable multivibrator circuit (suggested values: *R1, R2* = 1 kΩ *R3, R4* = 10 kΩ)

This latch circuit is similar to an astable multivibrator, except that there is no charge or discharge time, due to the absence of capacitors. Hence, when the circuit is switched on, if Q1 is on, its collector is at 0 V. As a result, Q2 gets switched off. This results in more than half $+V$ volts being applied to R4 causing current into the base of Q1, thus keeping it on. Thus, the circuit remains stable in a single state continuously. Similarly, Q2 remains on continuously, if it happens to get switched on first.

Switching of state can be done via Set and Reset terminals connected to the bases. For example, if Q2 is on and Set is grounded momentarily, this switches Q2 off, and makes Q1 on. Thus, Set is used to "set" Q1 on, and Reset is used to "reset" it to off state.

REFERENCES

1. US Patent 7132883 - Chopper chopper-stabilized instrumentation and operational amplifiers
2. University of New Orleans, Physics Department, Graduate student research.
3. A private laboratory in Cross Lanes, WV.
4. Bedford, B. D.; Hoft, R. G. (1964). Principles of Inverter Circuits. New York: John Wiley & Sons, Inc. ISBN 0-471-06134-4.
5. Fundamental Principals of Vibrator Power Supply Design (PDF) (First ed.). Indianapolis, Indiana: P. R. Mallory & Co., Inc. March 1947. p. 9.
6. "Vibrator Power Supplies". www.radioremembered.org. Retrieved 2016-01-24.
7. "Radio Vibrators MikeHaganAntiqueAutoRadio.com". www.mikehaganantiqueautoradio.com. Retrieved 2016-01-24.
8. Jain, R. P.; Anand, M. M. S. (1983). Digital Electronics Practice Using Integrated Circuits. Tata McGraw-Hill Education. p. 159. ISBN 0074516922.
9. Rao, Prakash (2006). Pulse And Digital Circuits. Tata McGraw-Hill Education. p. 268. ISBN 0070606560.
10. Clayton, G B (2013). Operational Amplifiers, 2nd Ed. Elsevier. p. 267. ISBN 1483135551.
11. Abraham, H.; E. Bloch (1919). "Mesure en valeur absolue des périodes des oscillations électriques de haute fréquence» [Measurement of the periods of high frequency electrical oscillations]. Annales de Physique (in French). Paris: Société Française de Physique. 9 (1): 237–302. doi:10.1051/jphystap:01919009002100.
12. Ginoux, Jean-Marc (2012). «Van der Pol and the history of relaxation oscillations: Toward the emergence of a concepts». Chaos 22 (2012) 023120. doi:10.1063/1.3670008.
13. Multivibrator in IEEE Std. 100 Dictionary of Standards Terms 7th ed.,IEEE Press, 2000 ISBN 0-7381-2601-2 page 718
14. William Henry Eccles and Frank Wilfred Jordan, «Improvements in ionic relays» British patent number: GB 148582 (filed: 21 June 1918; published: 5 August 1920).

15. Wilfred Bennett Lewis (1942). Electrical counting: with special reference to counting alpha and beta particles. CUP Archive. p. 68.
16. The Electrician. 128. Feb 13, 1942. Missing or empty |title= (help)
17. Owen Standige Puckle and E. B. Moullin (1943). Time bases (scanning generators): their design and development, with notes on the cathode ray tube. Chapman & Hall Ltd. p. 51.
18. Britton Chance (1949). Waveforms (Vol. 19 of MIT Radiation Lab Series ed.). McGraw-Hill Book Co. p. 167.
19. O. S. Puckle (Jan 1949). «Development of Time Bases: The Principles of Known Circuits». Wireless Engineer. Iliffe Electrical Publications. 26 (1): 139.
20. Donald Fink (ed), Electronics Engineers› Handbook, McGraw Hill, 1975 ISBN 0-07-020980-4, page 16-40

CHAPTER 4

DC-to-DC Converters

Chapter Contents

- **Introduction**
- **History**
- **Uses**
- **Electronic Conversion**
- **Electromechanical Conversion**
- **Electrochemical Conversion**

A DC-DC converter is a type of electronic circuit that takes the DC source and then converts it from the current voltage into another voltage. In some instances, one may have to increase or decrease the voltage. This is actually very common when it comes to automobiles. Some devices you might use in your vehicle, such as a portable DVD player for the kids, require a certain amount of power to run. Too much power and it could destroy the devices. Too little power and it would not run. Since the device needs to draw from the vehicle's battery, and the battery might be too powerful, there needs to be a conversion of power. However, it needs to remain DC voltage. To accomplish this, a variety of different types of DC-DC converters is available.

The converters will actually take the power from the battery and cut it down to a usable level. In the cases of devices that might need more power, a converter can step up the power. Many of the devices that you use in the vehicle come with adapters that you can plug into the vehicle. The adapters will convert the power to ensure your device works appropriately. It is important to make sure you are always using the right adapter with the device though. Otherwise, it could be drawing too much power, which could seriously damage the electronics.

INTRODUCTION

A DC-to-DC converter is an electronic circuit or electromechanical device that converts a source of direct current (DC) from one voltage level to another. It is a type of electric power converter. Power levels range from very low (small batteries) to very high (high-voltage power transmission).

HISTORY

Before the development of power semiconductors and allied technologies, one way to convert the voltage of a DC supply to a higher voltage, for low-power applications, was to convert it to AC by using a vibrator, followed by a step-up transformer and rectifier. For higher power an electric motor was used to drive a generator of the desired voltage (sometimes combined into a single "dynamotor" unit, a motor and generator combined into one unit, with one winding driving the motor and the other generating the output voltage). These were relatively inefficient and expensive procedures used only when there was no alternative, as to power a car radio (which then used thermionic valves/tubes requiring much higher voltages than available from a 6 or 12 V car battery). The introduction of power semiconductors and integrated circuits made it economically viable to use techniques as described below, for example to convert the DC power supply to high-frequency AC, use a transformer—small, light, and cheap due to the high frequency—to change the voltage, and rectify back to DC. Although by 1976 transistor car radio receivers did not require high voltages, some amateur radio operators continued to use vibrator supplies and dynamotors for mobile transceivers requiring high voltages, although transistorized power supplies were available.

While it was possible to derive a *lower* voltage from a higher with a linear electronic circuit, or even a resistor, these methods dissipated the excess as heat; energy-efficient conversion only became possible with solid-state switch-mode circuits.

USES

DC to DC converters are used in portable electronic devices such as cellular phones and laptop computers, which are supplied with power from batteries primarily. Such electronic devices often contain several sub-circuits, each with its own voltage level requirement different from that supplied by the battery or an external supply (sometimes higher or lower than the supply voltage). Additionally, the battery voltage declines as its stored energy is drained. Switched DC to DC converters offer a method to increase voltage from a partially lowered battery voltage thereby saving space instead of using multiple batteries to accomplish the same thing.

Most DC to DC converter circuits also regulate the output voltage. Some exceptions include high-efficiency LED power sources, which are a kind of DC to DC converter that regulates the current through the LEDs, and simple charge pumps which double or triple the output voltage.

DC to DC converters developed to maximize the energy harvest for photovoltaic systems and for wind turbines are called power optimizers.

Transformers used for voltage conversion at mains frequencies of 50–60 Hz must be large and heavy for powers exceeding a few watts. This makes them expensive, and they are subject to energy losses in their windings and due to eddy currents in their cores. DC-to-DC techniques that use transformers or inductors work at much higher frequencies, requiring only much smaller, lighter, and cheaper wound components. Consequently these techniques are used even where a mains transformer could be used; for example, for domestic electronic appliances it is preferable to rectify mains voltage to DC, use switch-mode techniques to convert it to high-frequency AC at the desired voltage, then, usually, rectify to DC. The entire complex circuit is cheaper and more efficient than a simple mains transformer circuit of the same output.

ELECTRONIC CONVERSION

Linear regulators which are used to output a stable DC independent of input voltage and output load from a higher but less stable input by dissipating excess volt-amperes as heat, could be described literally as DC-to-DC converters, but this is not usual usage. (The same could be said of a simple voltage dropper resistor, whether or not stabilised by a following voltage regulator or Zener diode.)

There are also simple capacitive voltage doubler and Dickson multiplier circuits using diodes and capacitors to multiply a DC voltage by an integer value, typically delivering only a small current.

Practical electronic converters use switching techniques. Switched-mode DC-to-DC converters convert one DC voltage level to another, which may be higher or lower, by storing the input energy temporarily and then releasing that energy to the output at a different voltage. The storage may be in either magnetic field storage components (inductors, transformers) or electric field storage components (capacitors). This conversion method can increase or decrease voltage. Switching conversion is more power efficient (often 75% to 98%) than linear voltage regulation, which dissipates unwanted power as heat. Fast semiconductor device rise and fall times are required for efficiency; however, these fast transitions combine with layout parasitic effects to make circuit design challenging. The higher efficiency of a switched-mode converter reduces the heatsinking needed, and increases battery endurance of portable equipment. Efficiency has improved since the late 1980s due to the use of power FETs, which are able to switch more efficiently with lower switching losses at higher frequencies than power bipolar transistors, and use less complex drive circuitry. Another important improvement in DC-DC converters is replacing the flywheel diode by synchronous rectification using a power FET, whose "on resistance" is much lower, reducing switching losses. Before the wide availability of power semiconductors, low-power DC-to-DC synchronous converters consisted of an electro-mechanical vibrator followed by a voltage step-up transformer feeding a vacuum tube or semiconductor rectifier, or synchronous rectifier contacts on the vibrator.

Most DC-to-DC converters are designed to move power in only one direction, from dedicated input to output. However, all switching regulator topologies can be made bidirectional and able to move power in either direction by replacing all diodes with independently controlled active rectification. A bidirectional converter is useful, for example, in applications requiring regenerative braking of vehicles, where power is supplied *to* the wheels while driving, but supplied *by* the wheels when braking.

Switching converters are electronically complex, although this is embodied in integrated circuits, with few components needed. They need careful design of the circuit and physical layout to reduce switching noise (EMI / RFI) to acceptable levels and, like all high-frequency circuits, for stable operation. Cost was higher than linear regulators in voltage-dropping applications, but this dropped with advances in chip design.

DC-to-DC converters are available as integrated circuits (ICs) requiring few additional components. Converters are also available as complete hybrid circuit modules, ready for use within an electronic assembly.

Magnetic

In these DC-to-DC converters, energy is periodically stored within and released from a magnetic field in an inductor or a transformer, typically within a frequency range of 300 kHz to 10 MHz. By adjusting the duty cycle of the charging voltage (that is, the ratio of the on/off times), the amount of power transferred to a load can be more easily controlled, though this control can also be applied to the input current, the output current, or to maintain constant power. Transformer-based converters may provide isolation between input and output. In general, the term *DC-to-DC converter* refers to one of these switching converters. These circuits are the heart of a switched-mode power supply. Many topologies exist. This table shows the most common ones.

<table>
<tr><th></th><th>Forward (energy transfers through the magnetic field)</th><th>Flyback (energy is stored in the magnetic field)</th></tr>
<tr><td rowspan="3">No transformer (non-isolated)</td><td>• Step-down (buck) - The output voltage is lower than the input voltage, and of the same polarity.</td><td>• Non-inverting: The output voltage is the same polarity as the input.
○ Step-up (boost) - The output voltage is higher than the input voltage.
○ SEPIC - The output voltage can be lower or higher than the input.
• Inverting: the output voltage is of the opposite polarity as the input.
○ Inverting (buck-boost).
○ Ćuk - Output current is continuous.</td></tr>
<tr><td colspan="2">• True buck-boost - The output voltage is the same polarity as the input and can be lower or higher.</td></tr>
<tr><td colspan="2">• Split-pi (boost-buck) - Allows bidirectional voltage conversion with the output voltage the same polarity as the input and can be lower or higher.</td></tr>
</table>

With transformer (isolatable)	• Forward - 1 or 2 transistor drive. • Push-pull (half bridge) - 2 transistors drive. • Full bridge - 4 transistor drive.	• Flyback - 1 transistor drive.

In addition, each topology may be:

Hard Switched

Transistors switch quickly while exposed to both full voltage and full current

Resonant

An LC circuit shapes the voltage across the transistor and current through it so that the transistor switches when either the voltage or the current is zero

Magnetic DC-to-DC converters may be operated in two modes, according to the current in its main magnetic component (inductor or transformer):

Continuous

The current fluctuates but never goes down to zero

Discontinuous

The current fluctuates during the cycle, going down to zero at or before the end of each cycle

A converter may be designed to operate in continuous mode at high power, and in discontinuous mode at low power.

The half bridge and flyback topologies are similar in that energy stored in the magnetic core needs to be dissipated so that the core does not saturate. Power transmission in a flyback circuit is limited by the amount of energy that can be stored in the core, while forward circuits are usually limited by the I/V characteristics of the switches.

Although MOSFET switches can tolerate simultaneous full current and voltage (although thermal stress and electromigration can shorten the MTBF), bipolar switches generally can't so require the use of a snubber (or two).

High-current systems often use multiphase converters, also called interleaved converters. Multiphase regulators can have better ripple and better response times than single-phase regulators.

Many laptop and desktop motherboards include interleaved buck regulators, sometimes as a voltage regulator module.

Capacitive

Switched capacitor converters rely on alternately connecting capacitors to the input and output in differing topologies. For example, a switched-capacitor reducing converter might charge two capacitors in series and then discharge them in parallel. This would produce the same output power (less that lost to efficiency of under 100%) at, ideally, half the input voltage and twice the current. Because they operate on discrete quantities of charge, these are also sometimes referred to as charge pump converters. They are typically used in applications requiring relatively small currents, as at higher currents the increased efficiency and smaller size of switch-mode converters makes them a better choice. They are also used at extremely high voltages, as magnetics would break down at such voltages.

ELECTROMECHANICAL CONVERSION

A motor-generator set, mainly of historical interest, consists of an electric motor and generator coupled together. A *dynamotor* combines both functions into a single unit with coils for both the motor and the generator functions wound around a single rotor; both coils share the same outer field coils or magnets. Typically the motor coils are driven from a commutator on one end of the shaft, when the generator coils output to another commutator on the other end of the shaft. The entire rotor and shaft assembly is smaller in size than a pair of machines, and may not have any exposed drive shafts.

A motor generator with separate motor and generator.

Motor-generators can convert between any combination of DC and AC voltage and phase standards. Large motor-generator sets were widely used to convert industrial amounts of power while smaller units were used to convert battery power (6, 12 or 24 V DC) to a high DC voltage, which was required to operate vacuum tube (thermionic valve) equipment.

For lower-power requirements at voltages higher than supplied by a vehicle battery, vibrator or "buzzer" power supplies were used. The vibrator oscillated mechanically, with contacts that switched the polarity of the battery many times per second, effectively converting DC to square wave AC, which could then be fed to a transformer of the required output voltage(s). It made a characteristic buzzing noise.

ELECTROCHEMICAL CONVERSION

A further means of DC to DC conversion in the kilowatts to megawatts range is presented by using redox flow batteries such as the vanadium redox battery.

Chaotic Behavior

DC-to-DC converters are subject to different types of chaotic dynamics such as bifurcation, crisis, and intermittency.

Terminology

Step-Down

A converter where output voltage is lower than the input voltage (such as a buck converter).

Step-Up

A converter that outputs a voltage higher than the input voltage (such as a boost converter).

Continuous Current Mode

Current and thus the magnetic field in the inductive energy storage never reach zero.

Discontinuous Current Mode

Current and thus the magnetic field in the inductive energy storage may reach or cross zero.

Noise

Unwanted electrical and electromagnetic signal noise, typically switching artifacts.

RF Noise

Switching converters inherently emit radio waves at the switching frequency and its harmonics. Switching converters that produce triangular switching current, such as the Split-Pi, forward converter, or Ćuk converter in continuous current mode, produce less harmonic noise than other switching converters. RF noise causes electromagnetic interference (EMI). Acceptable levels depend upon requirements, e.g. proximity to RF circuitry needs more suppression than simply meeting regulations.

Input Noise

The input voltage may have non-negligible noise. Additionally, if the converter loads the input with sharp load edges, the converter can emit RF noise from the supplying power lines. This should be prevented with proper filtering in the input stage of the converter.

Output Noise

The output of an ideal DC-to-DC converter is a flat, constant output voltage. However, real converters produce a DC output upon which is superimposed some level of electrical noise. Switching converters produce switching noise at the switching frequency and its harmonics. Additionally, all electronic circuits have some thermal noise. Some sensitive radio-frequency and analog circuits require a power supply with so little noise that it can only be provided by a linear regulator. Some analog circuits which require a power supply with relatively low noise can tolerate some of the less-noisy switching converters, e.g. using continuous triangular waveforms rather than square waves.

REFERENCES

1. "Vibrator Power Supplies". Radioremembered.org. Retrieved 18 January 2016.
2. Ed Brorein (2012-05-16). "Watt's Up?: What Is Old is New Again: Soft-Switching and Synchronous Rectification in Vintage Automobile Radios". Keysight Technologies: Watt's Up?. Retrieved 2016-01-19.
3. There is at least one example of a very large (three refrigerator-sized cabinets) and complex pre-transistor switching regulator using thyratron gas-filled tubes, although they appear to be used as regulators rather than for DC-to-DC conversion as such. This was the 1958 power supply for the IBM 704 computer, using 90 kW of power.
4. Radio Amateur's Handbook 1976, pub. ARRL, p331-332
5. Andy Howard (2015-08-25). "How to Design DC-to-DC Converters". YouTube. Retrieved 2015-10-02.
6. Stephen Sangwine (2 March 2007). Electronic Components and Technology, Third Edition. CRC Press. p. 73. ISBN 978-1-4200-0768-8.
7. Jeff Barrow of Integrated Device Technology, Inc. (21 November 2011). "Understand and reduce DC/DC switching-converter ground noise". Eetimes.com. Retrieved 18 January 2016.
8. Damian Giaouris et al. "Foldings and grazings of tori in current controlled interleaved boost converters". doi: 10.1002/cta.1906.
9. Ron Crews and Kim Nielson. "Interleaving is Good for Boost Converters, Too". 2008.
10. Keith Billings. "Advantages of Interleaving Converters". 2003.
11. John Gallagher "Coupled Inductors Improve Multiphase Buck Efficiency". 2006.

12. Juliana Gjanci. "On-Chip Voltage Regulation for Power Management inSystem-on-Chip". 2006. p. 22-23.
13. Majumder, Ritwik; Ghosh, Arindam; Ledwich, Gerard F.; Zare, Firuz (2008). "Control of Parallel Converters for Load Sharing with Seamless Transfer between Grid Connected and Islanded Modes". eprints.qut.edu.au. Retrieved 2016-01-19.
14. Tse, Chi K.; Bernardo, Mario Di (2002). Complex behavior in switching power converters. Proceedings of the IEEE. pp. 768–781.
15. Iqbal, Sajid; et al. (2014). Study of bifurcation and chaos in dc-dc boost converter using discrete-time map. IEEE International Conference on Mechatronics and Control (ICMC'2014) 2014. doi:10.1109/ICMC.2014.7231874.
16. Fossas, , Enric; Olivar, Gerard (1996). "Study of chaos in the buck converter". Circuits and Systems I: Fundamental Theory and Applications, IEEE Transactions on: 13–25.
17. Making -5V 14-bit Quiet, section of Linear Technology Application Note 84, Kevin Hoskins, 1997, pp 57-59

CHAPTER 5

DC to AC Convertors

Chapter Contents

- Introduction
- History
- Uses
- Electronic Conversion
- Electromechanical Conversion
- Electrochemical Conversion

Traditionally generation of electricity has involved rotating machines to produce alternating sinusoidal voltage and current (A.C. systems). With the development of power electronics during the last several decades, semi-conductor devices are now frequently used to convert direct current (D.C.) to A.C. to power conventional alternating current systems.

Devices used to convert D.C. to A.C. are called inverters. These are used numerous applications, including PV systems, battery storage systems, traction drives, variable speed drives and others. While they are said to convert D.C. to A.C., this is a simplification, and as we will see, the output of any inverter is built up by pulsed D.C. voltages.

INTRODUCTION

A power inverter, or inverter, is an electronic device or circuitry that changes direct current (DC) to alternating current (AC).

The input voltage, output voltage and frequency, and overall power handling depend on the design of the specific device or circuitry. The inverter does not produce any power; the power is provided by the DC source.

A power inverter can be entirely electronic or may be a combination of mechanical effects (such as a rotary apparatus) and electronic circuitry. Static inverters do not use moving parts in the conversion process.

An inverter on a free-standing solar plant.

Overview of solar-plant inverters.

INPUT AND OUTPUT

Input Voltage

A typical power inverter device or circuit requires a relatively stable DC power source capable of supplying enough current for the intended power demands of the system. The input voltage depends on the design and purpose of the inverter. Examples include:

- 12 VDC, for smaller consumer and commercial inverters that typically run from a rechargeable 12 V lead acid battery or automotive electrical outlet.
- 24, 36 and 48 VDC, which are common standards for home energy systems.
- 200 to 400 VDC, when power is from photovoltaic solar panels.
- 300 to 450 VDC, when power is from electric vehicle battery packs in vehicle-to-grid systems.
- Hundreds of thousands of volts, where the inverter is part of a high voltage direct current power transmission system.

Output Waveform

An inverter can produce a square wave, modified sine wave, pulsed sine wave, pulse width modulated wave (PWM) or sine wave depending on circuit design. The two dominant commercialized waveform types of inverters as of 2007 are modified sine wave and sine wave.

There are two basic designs for producing household plug-in voltage from a lower-voltage DC source, the first of which uses a switching boost converter to produce a higher-voltage DC and then converts to AC. The second method converts DC to AC at battery level and uses a line-frequency transformer to create the output voltage.

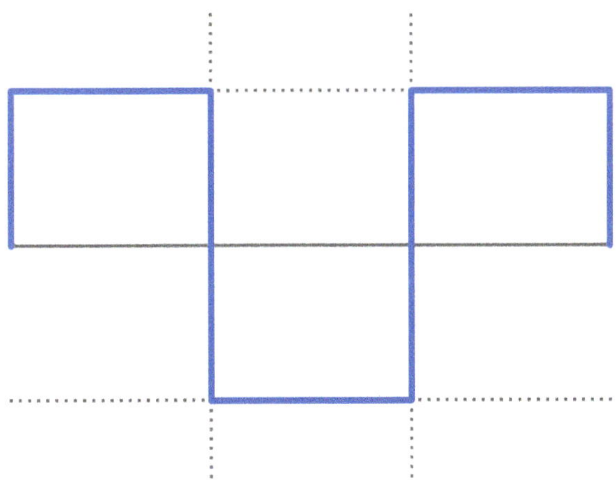

Square wave.

Square Wave

This is one of the simplest waveforms an inverter design can produce and is best suited to low-sensitivity applications such as lighting and heating. Square wave output can produce "humming" when connected to audio equipment and is generally unsuitable for sensitive electronics.

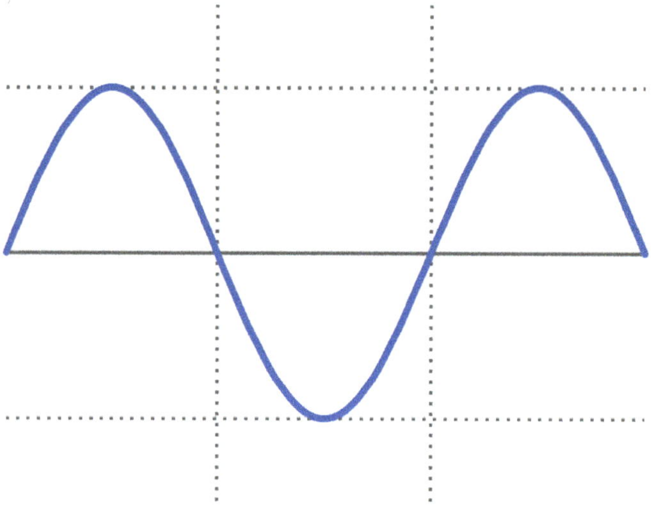

Sine wave.

Sine Wave

A power inverter device which produces a multiple step sinusoidal AC waveform is referred to as a *sine wave inverter*. To more clearly distinguish the inverters with outputs of much less distortion than the *modified sine wave* (three step) inverter designs, the manufacturers often use the phrase *pure sine wave inverter*. Almost all consumer grade inverters that are sold as a "pure sine wave inverter" do not produce a smooth sine wave output at all, just a less choppy output than the square wave (two step) and modified sine wave (three step) inverters. However, this is not critical for most electronics as they deal with the output quite well.

Where power inverter devices substitute for standard line power, a sine wave output is desirable because many electrical products are engineered to work best with a sine wave AC power source. The standard electric utility provides a sine wave, typically with minor imperfections but sometimes with significant distortion.

Sine wave inverters with more than three steps in the wave output are more complex and have significantly higher cost than a modified sine wave, with only three steps, or square wave (one step) types of the same power handling. Switch-mode power supply (SMPS) devices, such as personal computers or DVD players, function on quality modified sine wave power. AC motors directly operated on non-sinusoidal power may produce extra heat, may have different speed-torque characteristics, or may produce more audible noise than when running on sinusoidal power.

Modified Sine Wave

The modified sine wave output of such an inverter is the sum of two square waves one of which is phase shifted 90 degrees relative to the other. The result is three level waveform with equal intervals of zero volts; peak positive volts; zero volts; peak negative volts and then zero volts. This sequence is repeated. The resultant wave very roughly resembles the shape of a sine wave. Most inexpensive consumer power inverters produce a modified sine wave rather than a pure sine wave.

The waveform in commercially available modified-sine-wave inverters resembles a square wave but with a pause during the polarity reversal. Switching states are developed for positive, negative and zero voltages. The peak voltage to RMS voltage ratio does not maintain the same relationship as for a sine wave. The DC bus voltage may be actively regulated, or the "on" and "off" times can be modified to maintain the same RMS value output up to the DC bus voltage to compensate for DC bus voltage variations.

The ratio of on to off time can be adjusted to vary the RMS voltage while maintaining a constant frequency with a technique called pulse width modulation (PWM). The generated gate pulses are given to each switch in accordance with the developed pattern to obtain the desired output. Harmonic spectrum in the output depends on the width of the pulses and the modulation frequency. When operating induction motors, voltage harmonics are usually not of concern; however, harmonic distortion in the current waveform introduces additional heating and can produce pulsating torques.

Numerous items of electric equipment will operate quite well on modified sine wave power inverter devices, especially loads that are resistive in nature such as traditional incandescent light bulbs. Items with a switch-mode power supply operate almost entirely without problems, but if the item has a mains transformer, this can overheat depending on how marginally it is rated.

However, the load may operate less efficiently owing to the harmonics associated with a modified sine wave and produce a humming noise during operation. This also affects the efficiency of the system as a whole, since the manufacturer's nominal conversion efficiency does not account for harmonics. Therefore, pure sine wave inverters may provide significantly higher efficiency than modified sine wave inverters.

Most AC motors will run on MSW inverters with an efficiency reduction of about 20% owing to the harmonic content. However, they may be quite noisy. A series LC filter tuned to the fundamental frequency may help.

A common modified sine wave inverter topology found in consumer power inverters is as follows: An onboard microcontroller rapidly switches on and off power MOSFETs at high frequency like ~50 kHz. The MOSFETs directly pull from a low voltage DC source (such as a battery). This signal then goes through step-up transformers (generally many smaller transformers are placed in parallel to reduce the overall size of the inverter) to produce a higher voltage signal. The output of the step-up transformers then gets filtered by capacitors to produce a high voltage DC supply. Finally, this DC supply is pulsed with additional power MOSFETs by the microcontroller to produce the final modified sine wave signal.

Other Waveforms

By definition there is no restriction on the type of AC waveform an inverter might produce that would find use in a specific or special application.

Output Frequency

The AC output frequency of a power inverter device is usually the same as standard power line frequency, 50 or 60 hertz

If the output of the device or circuit is to be further conditioned (for example stepped up) then the frequency may be much higher for good transformer efficiency.

Output Voltage

The AC output voltage of a power inverter is often regulated to be the same as the grid line voltage, typically 120 or 240 VAC at the distribution level, even when there are changes in the load that the inverter is driving. This allows the inverter to power numerous devices designed for standard line power.

Some inverters also allow selectable or continuously variable output voltages.

Output Power

A power inverter will often have an overall power rating expressed in watts or kilowatts. This describes the power that will be available to the device the inverter is driving and, indirectly, the power that will be needed from the DC source. Smaller popular consumer and commercial devices designed to mimic line power typically range from 150 to 3000 watts.

Not all inverter applications are solely or primarily concerned with power delivery; in some cases the frequency and or waveform properties are used by the follow-on circuit or device.

Batteries

The runtime of an inverter is dependent on the battery power and the amount of power being drawn from the inverter at a given time. As the amount of equipment using the inverter increases, the runtime will decrease. In order to prolong the runtime of an inverter, additional batteries can be added to the inverter.

When attempting to add more batteries to an inverter, there are two basic options for installation:

Series Configuration

If the goal is to increase the overall voltage of the inverter, one can daisy chain batteries in a series configuration. In a series configuration, if a single battery dies, the other batteries will not be able to power the load.

Parallel Configuration

If the goal is to increase capacity and prolong the runtime of the inverter, batteries can be connected in parallel. This increases the overall ampere-hour (Ah) rating of the battery set.

If a single battery is discharged though, the other batteries will then discharge through it. This can lead to rapid discharge of the entire pack, or even an over-current and possible fire. To avoid this, large paralleled batteries may be connected via diodes or intelligent monitoring with automatic switching to isolate an under-voltage battery from the others.

APPLICATIONS

DC Power Source Usage

Inverter designed to provide 115 VAC from the 12 VDC source provided in an automobile. The unit shown provides up to 1.2 amperes of alternating current, or enough to power two sixty watt light bulbs.

An inverter converts the DC electricity from sources such as batteries or fuel cells to AC electricity. The electricity can be at any required voltage; in particular it can operate AC equipment designed for mains operation, or rectified to produce DC at any desired voltage.

Uninterruptible Power Supplies

An uninterruptible power supply (UPS) uses batteries and an inverter to supply AC power when mains power is not available. When mains power is restored, a rectifier supplies DC power to recharge the batteries.

Electric Motor Speed Control

Inverter circuits designed to produce a variable output voltage range are often used within motor speed controllers. The DC power for the inverter section can be derived from a normal AC wall outlet or some other source. Control and feedback circuitry is used to adjust the final output of the inverter section which will ultimately determine the speed of the motor operating under its mechanical load. Motor speed control needs are numerous and include things like: industrial motor driven equipment, electric vehicles, rail transport systems, and power tools. (See related: variable-frequency drive) Switching states are developed for positive, negative and zero voltages as per the patterns given in the switching Table 1.The generated gate pulses are given to each switch in accordance with the developed pattern and thus the output is obtained.

In Refrigeration Compressors

An inverter can be used to control the speed of the compressor motor to drive variable refrigerant flow in a refrigeration or air conditioning system to regulate system performance. Such installations are known as inverter compressors. Traditional methods of refrigeration regulation use single-speed compressors switched on and off periodically; inverter-equipped systems have a variable-frequency drive that control the speed of the motor and thus the compressor and cooling output. The variable-frequency AC from the inverter drives a brushless or induction motor, the speed of which is proportional to the frequency of the AC it is fed, so the compressor can be run at variable speeds—eliminating compressor stop-start cycles increases efficiency. A microcontroller typically monitors the temperature in the space to be cooled, and adjusts the speed of the compressor to maintain the desired temperature. The additional electronics and system hardware add cost to the equipment, but can result in substantial savings in operating costs.

Power Grid

Grid-tied inverters are designed to feed into the electric power distribution system. They transfer synchronously with the line and have as little harmonic content as possible. They also need a means of detecting the presence of utility power for safety reasons, so as not to continue to dangerously feed power to the grid during a power outage.

Solar

Internal view of a solar inverter. Note the many large capacitors (blue cylinders), used to store energy briefly and improve the output waveform.

A solar inverter is a balance of system (BOS) component of a photovoltaic system and can be used for both, grid-connected and off-grid systems. Solar inverters have special functions adapted for use with photovoltaic arrays, including maximum power point tracking and anti-islanding protection. Solar micro-inverters differ from conventional inverters, as an individual micro-inverter is attached to each solar panel. This can improve the overall efficiency of the system. The output from several micro-inverters is then combined and often fed to the electrical grid.

Induction Heating

Inverters convert low frequency main AC power to higher frequency for use in induction heating. To do this, AC power is first rectified to provide DC power. The inverter then changes the DC power to high frequency AC power. Due to the reduction in the number of DC sources employed, the structure becomes more reliable and the output voltage has higher resolution due to an increase in the number of steps so that the reference sinusoidal voltage can be better achieved. This configuration has recently become very popular in AC power supply and adjustable speed drive applications. This new inverter can avoid extra clamping diodes or voltage balancing capacitors.

There are three kinds of level shifted modulation techniques, namely:

- Phase Opposition Disposition (POD)
- Alternative Phase Opposition Disposition (APOD)

- Phase Disposition (PD)

HVDC Power Transmission

With HVDC power transmission, AC power is rectified and high voltage DC power is transmitted to another location. At the receiving location, an inverter in a static inverter plant converts the power back to AC. The inverter must be synchronized with grid frequency and phase and minimize harmonic generation.

Electroshock Weapons

Electroshock weapons and tasers have a DC/AC inverter to generate several tens of thousands of V AC out of a small 9 V DC battery. First the 9 V DC is converted to 400–2000 V AC with a compact high frequency transformer, which is then rectified and temporarily stored in a high voltage capacitor until a pre-set threshold voltage is reached. When the threshold (set by way of an airgap or TRIAC) is reached, the capacitor dumps its entire load into a pulse transformer which then steps it up to its final output voltage of 20–60 kV. A variant of the principle is also used in electronic flash and bug zappers, though they rely on a capacitor-based voltage multiplier to achieve their high voltage.

Miscellaneous

Typical applications for power inverters include:

- Portable consumer devices that allow the user to connect a battery, or set of batteries, to the device to produce AC power to run various electrical items such as lights, televisions, kitchen appliances, and power tools.
- Use in power generation systems such as electric utility companies or solar generating systems to convert DC power to AC power.
- Use within any larger electronic system where an engineering need exists for deriving an AC source from a DC source.

CIRCUIT DESCRIPTION

Basic Design

In one simple inverter circuit, DC power is connected to a transformer through the center tap of the primary winding. A switch is rapidly switched back and forth to allow current to flow back to the DC source following two alternate paths through one end of the primary winding and then the other. The alternation of the direction of current in the primary winding of the transformer produces alternating current (AC) in the secondary circuit.

The electromechanical version of the switching device includes two stationary contacts and a spring supported moving contact. The spring holds the movable contact against one of the stationary contacts and an electromagnet pulls the movable contact to the opposite stationary contact. The current in the electromagnet is interrupted by the action of the switch so that the switch continually switches rapidly back and forth. This type of electromechanical inverter switch, called a vibrator or buzzer, was once used in vacuum tube automobile radios. A similar mechanism has been used in door bells, buzzers and tattoo machines.

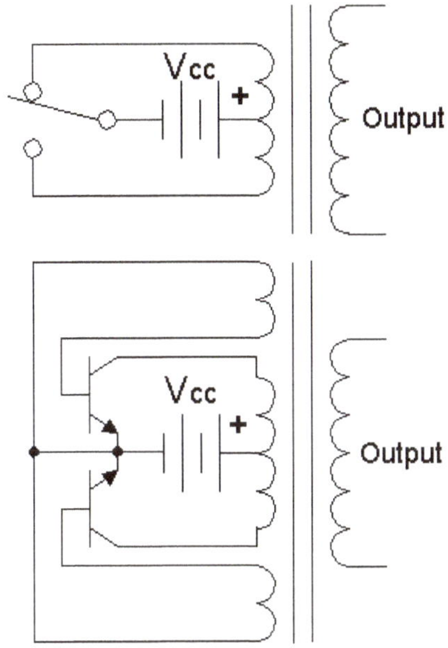

Top: Simple inverter circuit shown with an electromechanical switch *and automatic equivalent* auto-switching device implemented with two transistors and split winding auto-transformer in place of the mechanical switch.

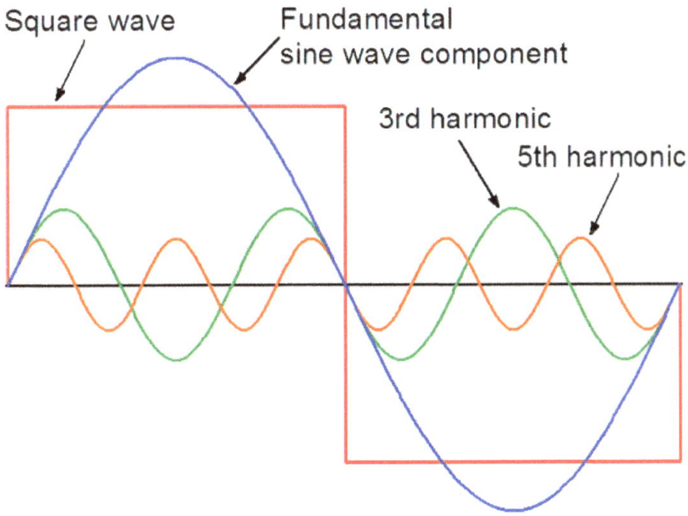

Square waveform with fundamental sine wave component, 3rd harmonic and 5th harmonic

As they became available with adequate power ratings, transistors and various other types of semiconductor switches have been incorporated into inverter circuit designs. Certain ratings, especially for large systems (many kilowatts) use thyristors (SCR). SCRs provide large power handling capability in a semiconductor device, and can readily be controlled over a variable firing range.

The switch in the simple inverter described above, when not coupled to an output transformer, produces a square voltage waveform due to its simple off and on nature as opposed to the sinusoidal waveform that is the usual waveform of an AC power supply. Using Fourier analysis, periodic waveforms are represented as the sum of an infinite series of sine waves. The sine wave that has the same frequency as the original waveform is called the fundamental component. The other sine waves, called *harmonics*, that are included in the series have frequencies that are integral multiples of the fundamental frequency.

Fourier analysis can be used to calculate the total harmonic distortion (THD). The total harmonic distortion (THD) is the square root of the sum of the squares of the harmonic voltages divided by the fundamental voltage:

$$\mathrm{THD} = \frac{\sqrt{V_2^2 + V_3^2 + V_4^2 + \cdots + V_n^2}}{V_1}$$

Advanced Designs

There are many different power circuit topologies and control strategies used in inverter designs. Different design approaches address various issues that may be more or less important depending on the way that the inverter is intended to be used.

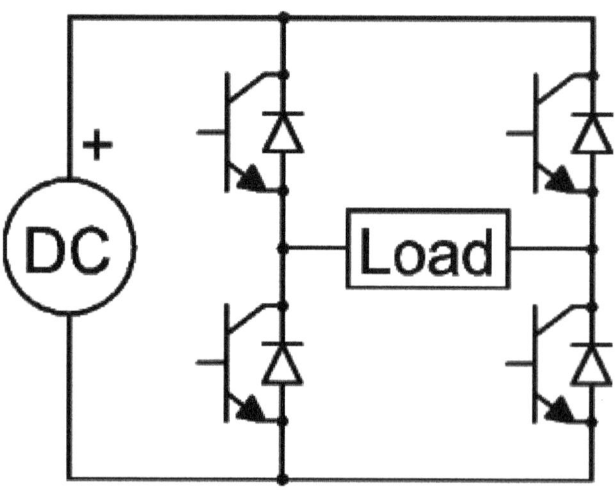

H bridge inverter circuit with transistor switches and antiparallel diodes

The issue of waveform quality can be addressed in many ways. Capacitors and inductors can be used to filter the waveform. If the design includes a transformer, filtering can be applied to the primary or the secondary side of the transformer or to both sides. Low-pass filters are applied to allow the fundamental component of the waveform to pass to the output while limiting the passage of the harmonic components. If the inverter is designed to provide power at a fixed frequency, a resonant filter can be used. For an adjustable frequency inverter, the filter must be tuned to a frequency that is above the maximum fundamental frequency.

Since most loads contain inductance, feedback rectifiers or antiparallel diodes are often connected across each semiconductor switch to provide a path for the peak inductive load current when the switch is turned off. The antiparallel diodes are somewhat similar to the *freewheeling diodes* used in AC/DC converter circuits.

Waveform	Signal transitions per period	Harmonics eliminated	Harmonics amplified	System description	THD
	2			2-level square wave	~45%
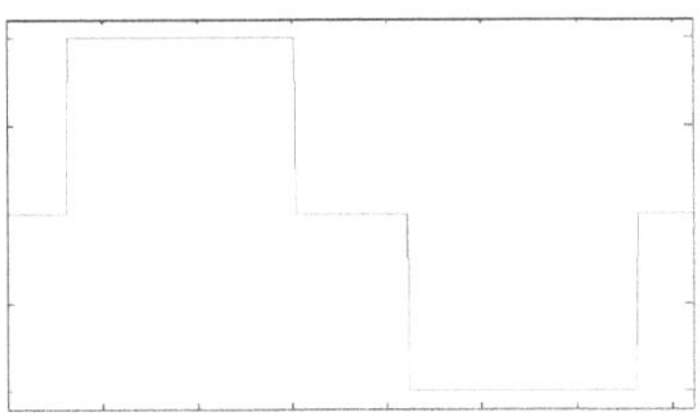	4	3, 9, 27, …		3-level modified sine wave	>23.8%
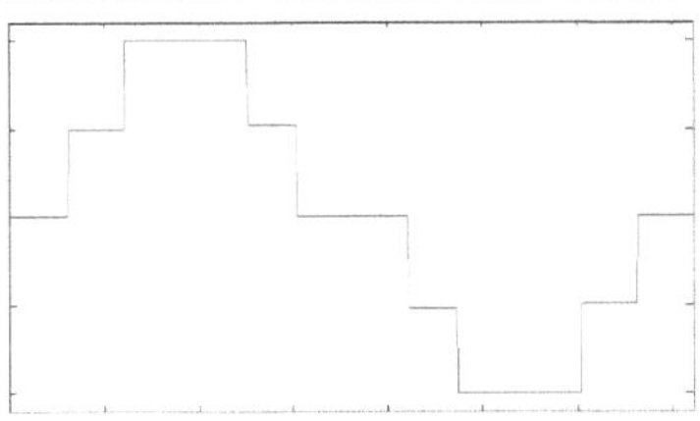	8			5-level modified sine wave	>6.5%
	10	3, 5, 9, 27	7, 11, …	2-level very slow PWM	

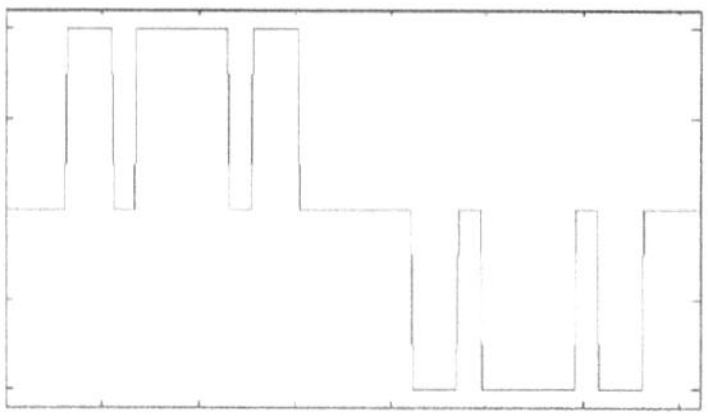	12	3, 5, 9, 27	7, 11, …	3-level very slow PWM	

Fourier analysis reveals that a waveform, like a square wave, that is anti-symmetrical about the 180 degree point contains only odd harmonics, the 3rd, 5th, 7th, etc. Waveforms that have steps of certain widths and heights can attenuate certain lower harmonics at the expense of amplifying higher harmonics. For example, by inserting a zero-voltage step between the positive and negative sections of the square-wave, all of the harmonics that are divisible by three (3rd and 9th, etc.) can be eliminated. That leaves only the 5th, 7th, 11th, 13th etc. The required width of the steps is one third of the period for each of the positive and negative steps and one sixth of the period for each of the zero-voltage steps.

Changing the square wave as described above is an example of pulse-width modulation (PWM). Modulating, or regulating the width of a square-wave pulse is often used as a method of regulating or adjusting an inverter's output voltage. When voltage control is not required, a fixed pulse width can be selected to reduce or eliminate selected harmonics. Harmonic elimination techniques are generally applied to the lowest harmonics because filtering is much more practical at high frequencies, where the filter components can be much smaller and less expensive. *Multiple pulse-width* or *carrier based* PWM control schemes produce waveforms that are composed of many narrow pulses. The frequency represented by the number of narrow pulses per second is called the *switching frequency* or *carrier frequency*. These control schemes are often used in variable-frequency motor control inverters because they allow a wide range of output voltage and frequency adjustment while also improving the quality of the waveform.

Multilevel inverters provide another approach to harmonic cancellation. Multilevel inverters provide an output waveform that exhibits multiple steps at several voltage levels. For example, it is possible to produce a more sinusoidal wave by having split-rail direct current inputs at two voltages, or positive and negative inputs with a central ground. By connecting the inverter output terminals in sequence between the positive rail and ground, the positive rail and the negative rail, the ground rail and the negative rail, then both to the ground rail, a stepped waveform is generated at the inverter output. This is an example of a three level inverter: the two voltages and ground.

More on Achieving a Sine Wave

Resonant inverters produce sine waves with LC circuits to remove the harmonics from a simple square wave. Typically there are several series- and parallel-resonant LC circuits, each tuned to a different harmonic of the power line frequency. This simplifies the electronics, but the inductors and capacitors tend to be large and heavy. Its high efficiency makes this approach popular in large uninterruptible power supplies in data centers that run the inverter continuously in an "online" mode to avoid any switchover transient when power is lost. (See related: Resonant inverter)

A closely related approach uses a ferroresonant transformer, also known as a constant voltage transformer, to remove harmonics and to store enough energy to sustain the load for a few AC cycles. This property makes them useful in standby power supplies to eliminate the switchover transient that otherwise occurs during a power failure while the normally idle inverter starts and the mechanical relays are switching to its output.

Enhanced Quantization

A proposal suggested in *Power Electronics* magazine utilizes two voltages as an improvement over the common commercialized technology, which can only apply DC bus voltage in either direction or turn it off. The proposal adds intermediate voltages to the common design. Each cycle sees the following sequence of delivered voltages: v1, v2, v1, 0, −v1, −v2, −v1.

Three-Phase Inverters

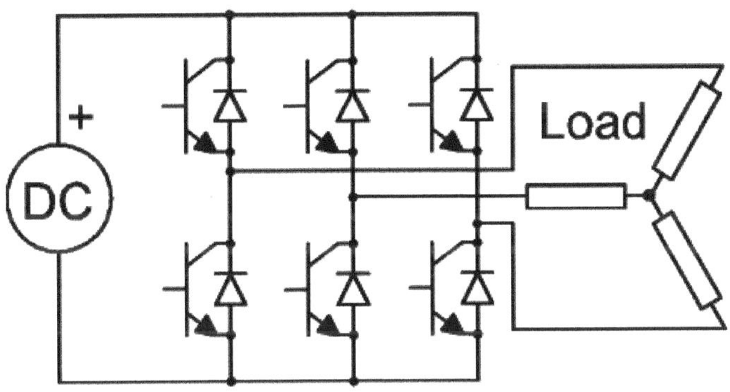

Three-phase inverter with wye connected load

Three-phase inverters are used for variable-frequency drive applications and for high power applications such as HVDC power transmission. A basic three-phase inverter consists of three single-phase inverter switches each connected to one of the three load terminals. For the most basic control scheme, the operation of the three switches is coordinated so that one switch operates at each 60 degree point of the fundamental output waveform. This creates a line-to-line output waveform that has six steps. The six-step waveform has a zero-voltage step between the positive and negative sections of the square-wave such that the harmonics that are multiples of three are eliminated as described above. When carrier-based PWM techniques are applied to six-step waveforms, the basic overall shape, or *envelope*, of the waveform is retained so that the 3rd harmonic and its multiples are cancelled.

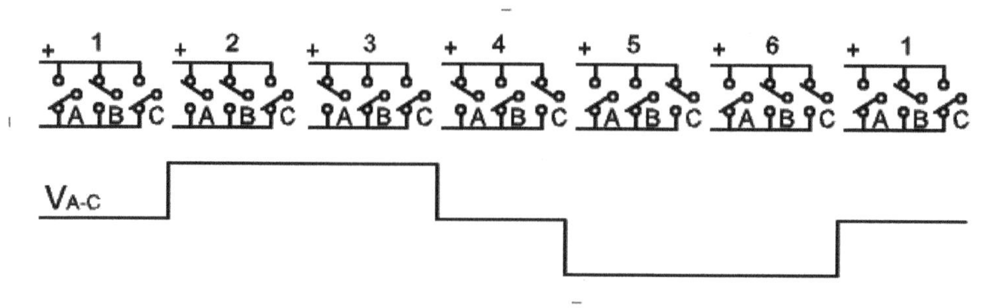

3-phase inverter switching circuit showing 6-step switching sequence and waveform of voltage between terminals A and C ($2^3 - 2$ states)

To construct inverters with higher power ratings, two six-step three-phase inverters can be connected in parallel for a higher current rating or in series for a higher voltage rating. In either case, the output waveforms are phase shifted to obtain a 12-step waveform. If additional inverters are combined, an 18-step inverter is obtained with three inverters etc. Although inverters are usually combined for the purpose of achieving increased voltage or current ratings, the quality of the waveform is improved as well.

Size

Compared to other household electric devices, inverters are large in size and volume. In 2014 Google together with IEEE started an open competition to build a (much) smaller power inverter, with a $1,000,000 prize.

History

Early Inverters

From the late nineteenth century through the middle of the twentieth century, DC-to-AC power conversion was accomplished using rotary converters or motor-generator sets (M-G sets). In the early twentieth century, vacuum tubes and gas filled tubes began to be used as switches in inverter circuits. The most widely used type of tube was the thyratron.

The origins of electromechanical inverters explain the source of the term *inverter*. Early AC-to-DC converters used an induction or synchronous AC motor direct-connected to a generator (dynamo) so that the generator's commutator reversed its connections at exactly the right moments to produce DC. A later development is the synchronous converter, in which the motor and generator windings are combined into one armature, with slip rings at one end and a commutator at the other and only one field frame. The result with either is AC-in, DC-out. With an M-G set, the DC can be considered to be separately generated from the AC; with a synchronous converter, in a certain sense it can be considered to be "mechanically rectified AC". Given the right auxiliary and control equipment, an M-G set or rotary converter can be "run backwards", converting DC to AC. Hence an inverter is an inverted converter.

Controlled Rectifier Inverters

Since early transistors were not available with sufficient voltage and current ratings for most inverter applications, it was the 1957 introduction of the thyristor or silicon-controlled rectifier (SCR) that initiated the transition to solid state inverter circuits.

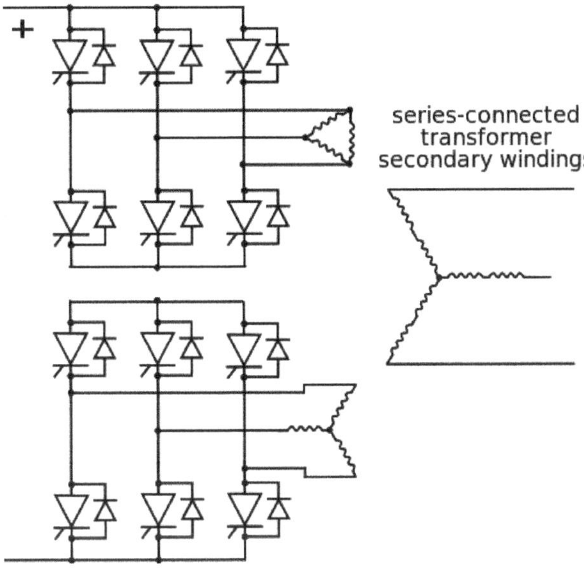

12-pulse line-commutated inverter circuit

The *commutation* requirements of SCRs are a key consideration in SCR circuit designs. SCRs do not turn off or *commutate* automatically when the gate control signal is shut off. They only turn off when the forward current is reduced to below the minimum holding current, which varies with each kind of SCR, through some external process. For SCRs connected to an AC power source, commutation occurs naturally every time the polarity of the source voltage reverses. SCRs connected to a DC power source usually require a means of forced commutation that forces the current to zero when commutation is required. The least complicated SCR circuits employ natural commutation rather than forced commutation. With the addition of forced commutation circuits, SCRs have been used in the types of inverter circuits described above.

In applications where inverters transfer power from a DC power source to an AC power source, it is possible to use AC-to-DC controlled rectifier circuits operating in the inversion mode. In the inversion mode, a controlled rectifier circuit operates as a line commutated inverter. This type of operation can be used in HVDC power transmission systems and in regenerative braking operation of motor control systems.

Another type of SCR inverter circuit is the current source input (CSI) inverter. A CSI inverter is the dual of a six-step voltage source inverter. With a current source inverter, the DC power supply is configured as a current source rather than a voltage source. The inverter SCRs are switched in a six-step sequence to direct the current to a three-phase AC load as a stepped current waveform. CSI inverter commutation methods include load commutation and parallel capacitor commutation. With both methods, the input current regulation assists the commutation. With load commutation, the load is a synchronous motor operated at a leading power factor.

As they have become available in higher voltage and current ratings, semiconductors such as transistors or IGBTs that can be turned off by means of control signals have become the preferred switching components for use in inverter circuits.

Rectifier and Inverter Pulse Numbers

Rectifier circuits are often classified by the number of current pulses that flow to the DC side of the rectifier per cycle of AC input voltage. A single-phase half-wave rectifier is a one-pulse circuit and a single-phase full-wave rectifier is a two-pulse circuit. A three-phase half-wave rectifier is a three-pulse circuit and a three-phase full-wave rectifier is a six-pulse circuit.

With three-phase rectifiers, two or more rectifiers are sometimes connected in series or parallel to obtain higher voltage or current ratings. The rectifier inputs are supplied from special transformers that provide phase shifted outputs. This has the effect of phase multiplication. Six phases are obtained from two transformers, twelve phases from three transformers and so on. The associated rectifier circuits are 12-pulse rectifiers, 18-pulse rectifiers and so on...

When controlled rectifier circuits are operated in the inversion mode, they would be classified by pulse number also. Rectifier circuits that have a higher pulse number have reduced harmonic content in the AC input current and reduced ripple in the DC output voltage. In the inversion mode, circuits that have a higher pulse number have lower harmonic content in the AC output voltage waveform.

Other Notes

The large switching devices for power transmission applications installed until 1970 predominantly used mercury-arc valves. Modern inverters are usually solid state (static inverters). A modern design method features components arranged in an H bridge configuration. This design is also quite popular with smaller-scale consumer devices.

Research

Using 3-D printing and novel semiconductors, researchers at the Department of Energy's Oak Ridge National Laboratory have created a power inverter that could make electric vehicles lighter, more powerful and more efficient.

UNINTERRUPTIBLE POWER SUPPLY

An uninterruptible power supply, also uninterruptible power source, UPS or battery/flywheel backup, is an electrical apparatus that provides emergency power to a load when the input power source or mains power fails. A UPS differs from an auxiliary or emergency power system or standby generator in that it will provide near-instantaneous protection from input power interruptions, by supplying energy stored in batteries, supercapacitors, or flywheels. The on-battery runtime of most uninterruptible power sources is relatively short (only a few minutes) but sufficient to start a standby power source or properly shut down the protected equipment.

A UPS is typically used to protect hardware such as computers, data centers, telecommunication equipment or other electrical equipment where an unexpected power disruption could cause injuries, fatalities, serious business disruption or data loss. UPS units range in size from units designed to protect a single computer without a video monitor (around 200 volt-ampere rating) to large units powering entire data centers or buildings. The world's largest UPS, the 46-megawatt Battery Electric Storage System (BESS), in Fairbanks, Alaska, powers the entire city and nearby rural communities during outages.

A small free-standing UPS with one IEC 60320 C14 input and three C13 outputs

A large datacenter-scale UPS being installed by electricians

Common Power Problems

The primary role of any UPS is to provide short-term power when the input power source fails. However, most UPS units are also capable in varying degrees of correcting common utility power problems:

- Voltage spike or sustained overvoltage
- Momentary or sustained reduction in input voltage
- Noise, defined as a high frequency transient or oscillation, usually injected into the line by nearby equipment
- Instability of the mains frequency
- Harmonic distortion, defined as a departure from the ideal sinusoidal waveform expected on the line

UPS units are divided into categories based on which of the above problems they address, and some manufacturers categorize their products in accordance with the number of power-related problems they address.

Technologies

The three general categories of modern UPS systems are *on-line*, *line-interactive* and *standby*. An on-line UPS uses a "double conversion" method of accepting AC input, rectifying to DC for passing through the rechargeable battery (or battery strings), then inverting back to 120 V/230 V AC for powering the protected equipment. A line-interactive UPS maintains the inverter in line and redirects the battery's DC current path from the normal charging mode to supplying current when power is lost. In a standby ("off-line") system the load is powered directly by the input power and the backup power circuitry is only invoked when the utility power fails. Most UPS below 1 kVA are of the line-interactive or standby variety which are usually less expensive.

For large power units, Dynamic Uninterruptible Power Supplies (DUPS) are sometimes used. A synchronous motor/alternator is connected on the mains via a choke. Energy is stored in a flywheel. When the mains power fails, an eddy-current regulation maintains the power on the load as long as the flywheel's energy is not exhausted. DUPS are sometimes combined or integrated with a diesel generator that is turned on after a brief delay, forming a diesel rotary uninterruptible power supply (DRUPS).

A fuel cell UPS has been developed in recent years using hydrogen and a fuel cell as a power source, potentially providing long run times in a small space.

Offline/Standby

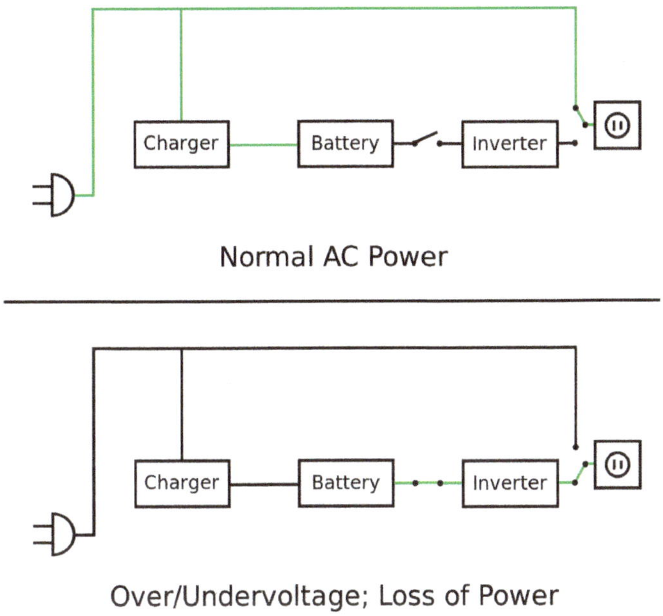

Offline/Standby UPS: The green line illustrates the flow of electric power. Typical protection time: 0–20 minutes. Capacity expansion: Usually not available

The offline/standby UPS (SPS) offers only the most basic features, providing surge protection and battery backup. The protected equipment is normally connected directly to incoming utility power. When the incoming voltage falls below or rises above a predetermined level the SPS turns on its internal DC-AC inverter circuitry, which is powered from an internal storage battery. The UPS then mechanically switches the connected equipment on to its DC-AC inverter output. The switchover time can be as long as 25 milliseconds depending on the amount of time it takes the standby UPS to detect the lost utility voltage. The UPS will be designed to power certain equipment, such as a personal computer, without any objectionable dip or brownout to that device.

Line-Interactive

The line-interactive UPS is similar in operation to a standby UPS, but with the addition of a multi-tap variable-voltage autotransformer. This is a special type of transformer that can add or subtract powered coils of wire, thereby increasing or decreasing the magnetic field and the output voltage of the transformer. This may also be performed by a *buck–boost transformer* which is distinct from an autotransformer, since the former may be wired to provide galvanic isolation.

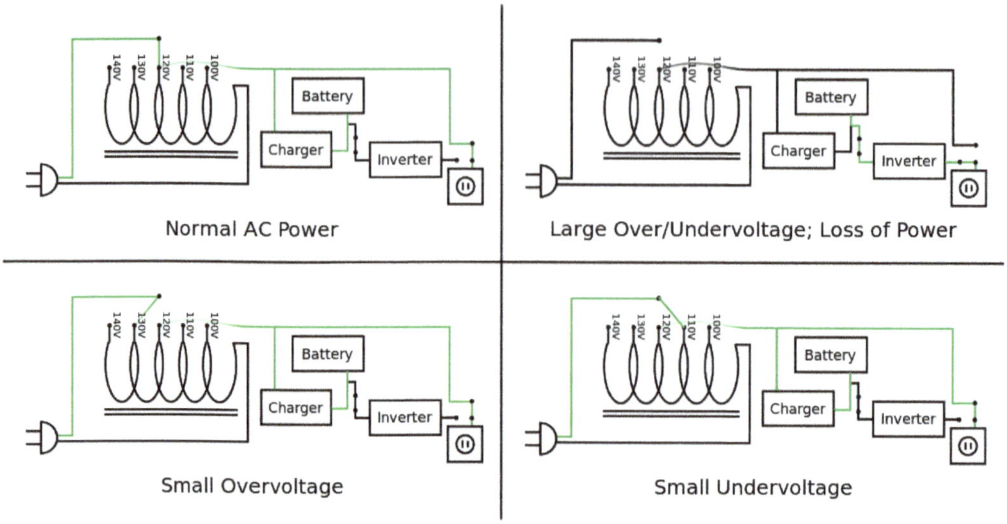

Line-interactive UPS: The green line illustrates the flow of electric power. Typical protection time: 5–30 minutes. Capacity expansion: several hours

This type of UPS is able to tolerate continuous undervoltage brownouts and overvoltage surges without consuming the limited reserve battery power. It instead compensates by automatically selecting different power taps on the autotransformer. Depending on the design, changing the autotransformer tap can cause a very brief output power disruption, which may cause UPSs equipped with a power-loss alarm to "chirp" for a moment.

This has become popular even in the cheapest UPSs because it takes advantage of components already included. The main 50/60 Hz transformer used to convert between line voltage and battery voltage needs to provide two slightly different turns ratios: One to convert the battery output voltage (typically a multiple of 12 V) to line voltage, and a second one to convert the line voltage to a slightly higher battery charging voltage (such as a multiple of 14 V). The difference between the two voltages is because charging a battery requires a delta voltage (up to 13–14 V for charging a 12 V battery). Furthermore, it is easier to do the switching on the line-voltage side of the transformer because of the lower currents on that side.

To gain the *buck/boost* feature, all that is required is two separate switches so that the AC input can be connected to one of the two primary taps, while the load is connected to the other, thus using the main transformer's primary windings as an autotransformer. The battery can still be charged while "bucking" an overvoltage, but while "boosting" an undervoltage, the transformer output is too low to charge the batteries.

Autotransformers can be engineered to cover a wide range of varying input voltages, but this requires more taps and increases complexity, and expense of the UPS. It is common for the autotransformer to cover a range only from about 90 V to 140 V for 120 V power, and then switch to battery if the voltage goes much higher or lower than that range.

In low-voltage conditions the UPS will use more current than normal so it may need a higher current circuit than a normal device. For example, to power a 1000-W device at 120 V, the UPS will draw 8.33 A. If a brownout occurs and the voltage drops to 100 V, the UPS will draw 10 A to compensate. This also works in reverse, so that in an overvoltage condition, the UPS will need less current.

Online/Double-Conversion

In an online UPS, the batteries are always connected to the inverter, so that no power transfer switches are necessary. When power loss occurs, the rectifier simply drops out of the circuit and the batteries keep the power steady and unchanged. When power is restored, the rectifier resumes carrying most of the load and begins charging the batteries, though the charging current may be limited to prevent the high-power rectifier from overheating the batteries and boiling off the electrolyte. The main advantage of an on-line UPS is its ability to provide an "electrical firewall" between the incoming utility power and sensitive electronic equipment.

The online UPS is ideal for environments where electrical isolation is necessary or for equipment that is very sensitive to power fluctuations. Although it was at one time reserved for very large installations of 10 kW or more, advances in technology have now permitted it to be available as a common consumer device, supplying 500 W or less. The initial cost of the online UPS may be higher, but its total cost of ownership is generally lower due to longer battery life. The online UPS may be necessary when the power environment is "noisy", when utility power sags, outages and other anomalies are frequent, when protection of sensitive IT equipment loads is required, or when operation from an extended-run backup generator is necessary.

The basic technology of the online UPS is the same as in a standby or line-interactive UPS. However it typically costs much more, due to it having a much greater current AC-to-DC battery-charger/rectifier, and with the rectifier and inverter designed to run continuously with improved cooling systems. It is called a *double-conversion* UPS due to the rectifier directly driving the inverter, even when powered from normal AC current.

Other Designs

Hybrid topology/Double Conversion on Demand

These hybrid Rotary UPS designs do not have official designations, although one name used by **UTL** is "double conversion on demand". This style of UPS is targeted towards high-efficiency applications while still maintaining the features and protection level offered by double conversion.

A hybrid (double conversion on demand) UPS operates as an off-line/standby UPS when power conditions are within a certain preset window. This allows the UPS to achieve very high efficiency ratings. When the power conditions fluctuate outside of the predefined windows, the UPS switches to online/double-conversion operation. In double-conversion mode the UPS can adjust for voltage variations without having to use battery power, can filter out line noise and control frequency. Examples of this hybrid/double conversion on demand UPS design are the HP R8000, HP R12000, HP RP12000/3 and the Eaton BladeUPS.

Ferro-Resonant

Ferro-resonant units operate in the same way as a standby UPS unit; however, they are online with the exception that a ferro-resonant transformer is used to filter the output. This transformer is designed to hold energy long enough to cover the time between switching from line power to battery power and effectively eliminates the transfer time. Many ferro-resonant UPSs are 82–88% efficient (AC/DC-AC) and offer excellent isolation.

The transformer has three windings, one for ordinary mains power, the second for rectified battery power, and the third for output AC power to the load.

This once was the dominant type of UPS and is limited to around the 150 kVA range. These units are still mainly used in some industrial settings (oil and gas, petrochemical, chemical, utility, and heavy industry markets) due to the robust nature of the UPS. Many ferro-resonant UPSs utilizing controlled ferro technology may not interact with power-factor-correcting equipment.

DC Power

A UPS designed for powering DC equipment is very similar to an online UPS, except that it does not need an output inverter. Also, if the UPS's battery voltage is matched with the voltage the device needs, the device's power supply will not be needed either. Since one or more power conversion steps are eliminated, this increases efficiency and run time.

Many systems used in telecommunications use an extra-low voltage "common battery" 48 V DC power, because it has less restrictive safety regulations, such as being installed in conduit and junction boxes. DC has typically been the dominant power source for telecommunications, and AC has typically been the dominant source for computers and servers.

There has been much experimentation with 48 V DC power for computer servers, in the hope of reducing the likelihood of failure and the cost of equipment. However, to supply the same amount of power, the current would be higher than an equivalent 115 V or 230 V circuit; greater current requires larger conductors, or more energy lost as heat.

A laptop computer is a classic example of a PC with a DC UPS built in.

High voltage DC (380 V) is finding use in some data center applications, and allows for small power conductors, but is subject to the more complex electrical code rules for safe containment of high voltages.

Rotary

A rotary UPS uses the inertia of a high-mass spinning flywheel (flywheel energy storage) to provide short-term *ride-through* in the event of power loss. The flywheel also acts as a buffer against power spikes and sags, since such short-term power events are not able to appreciably affect the rotational speed of the high-mass flywheel. It is also one of the oldest designs, predating vacuum tubes and integrated circuits.

It can be considered to be *on line* since it spins continuously under normal conditions. However, unlike a battery-based UPS, flywheel-based UPS systems typically provide 10 to 20 seconds of protection before the flywheel has slowed and power output stops. It is traditionally used in conjunction with standby diesel generators, providing backup power only for the brief period of time the engine needs to start running and stabilize its output.

The rotary UPS is generally reserved for applications needing more than 10,000 W of protection, to justify the expense and benefit from the advantages rotary UPS systems bring. A larger flywheel or multiple flywheels operating in parallel will increase the reserve running time or capacity.

Because the flywheels are a mechanical power source, it is not necessary to use an electric motor or generator as an intermediary between it and a diesel engine designed to provide emergency power. By using a transmission gearbox, the rotational inertia of the flywheel can be used to directly start up a diesel engine, and once running, the diesel engine can be used to directly spin the flywheel. Multiple flywheels can likewise be connected in parallel through mechanical countershafts, without the need for separate motors and generators for each flywheel.

They are normally designed to provide very high current output compared to a purely electronic UPS, and are better able to provide inrush current for inductive loads such as motor startup or compressor loads, as well as medical MRI and cath lab equipment. It is also able to tolerate short-circuit conditions up to 17 times larger than an electronic UPS, permitting one device to blow a fuse and fail while other devices still continue to be powered from the rotary UPS.

Its life cycle is usually far greater than a purely electronic UPS, up to 30 years or more. But they do require periodic downtime for mechanical maintenance, such as ball bearing replacement. In larger systems redundancy of the system ensures the availability of processes during this maintenance. Battery-based designs do not require downtime if the batteries can be hot-swapped, which is usually the case for larger units. Newer rotary units use technologies such as magnetic bearings and air-evacuated enclosures to increase standby efficiency and reduce maintenance to very low levels.

Typically, the high-mass flywheel is used in conjunction with a motor-generator system. These units can be configured as:

- A motor driving a mechanically connected generator,
- A combined synchronous motor and generator wound in alternating slots of a single rotor and stator,
- A hybrid rotary UPS, designed similar to an online UPS, except that it uses the flywheel in place of batteries. The rectifier drives a motor to spin the flywheel, while a generator uses the flywheel to power the inverter.

In case No. 3 the motor generator can be synchronous/synchronous or induction/synchronous. The motor side of the unit in case Nos. 2 and 3 can be driven directly by an AC power source (typically when in inverter bypass), a 6-step double-conversion motor drive, or a 6-pulse inverter. Case No. 1 uses an integrated flywheel as a short-term energy source instead of batteries to allow time for external, electrically coupled gensets to start and be brought online. Case Nos. 2 and 3 can use batteries or a free-standing electrically coupled flywheel as the short-term energy source.

Form Factors

UPS systems come in several different forms and sizes. However, the two most common forms are tower and rack-mount.

Tower Model

Tower models stand upright on the ground or on a desk/shelf, and are typically used in network workstations or desktop computer applications.

Rack-Mount Model

Rack-mount models can be mounted in standard 19" rack enclosures and can require anywhere from 1U to 12U (rack space). They are typically used in server and networking applications.

Applications

N+1

In large business environments where reliability is of great importance, a single huge UPS can also be a single point of failure that can disrupt many other systems. To provide greater reliability, multiple smaller UPS modules and batteries can be integrated together to provide redundant power protection equivalent to one very large UPS. "N+1" means that if the load can be supplied by N modules, the installation will contain N+1 modules. In this way, failure of one module will not impact system operation.

Multiple Redundancy

Many computer servers offer the option of redundant power supplies, so that in the event of one power supply failing, one or more other power supplies are able to power the load. This is a critical point – each power supply must be able to power the entire server by itself.

Redundancy is further enhanced by plugging each power supply into a different circuit (i.e. to a different circuit breaker).

Redundant protection can be extended further yet by connecting each power supply to its own UPS. This provides double protection from both a power supply failure and a UPS failure, so that continued operation is assured. This configuration is also referred to as 1+1 or 2N redundancy. If the budget does not allow for two identical UPS units then it is common practice to plug one power supply into mains power and the other into the UPS.

Outdoor Use

When a UPS system is placed outdoors, it should have some specific features that guarantee that it can tolerate weather without any effects on performance. Factors such as temperature, humidity, rain, and snow among others should be considered by the manufacturer when designing an outdoor UPS system. Operating temperature ranges for outdoor UPS systems could be around −40 °C to +55 °C.

Outdoor UPS systems can either be pole, ground (pedestal), or host mounted. Outdoor environment could mean extreme cold, in which case the outdoor UPS system should include a battery heater mat, or extreme heat, in which case the outdoor UPS system should include a fan system or an air conditioning system.

Internal view of a solar inverter. Note the many large capacitors (blue cylinders), used to store energy briefly and improve the output waveform.

A solar inverter, or PV inverter, or solar converter, converts the variable direct current (DC) output of a photovoltaic (PV) solar panel into a utility frequency alternating current (AC) that can be fed into a commercial electrical grid or used by a local, off-grid electrical network. It is a critical BOS–component in a photovoltaic system, allowing the use of ordinary AC-powered equipment. Solar inverters have special functions adapted for use with photovoltaic arrays, including maximum power point tracking and anti-islanding protection.

Difficulties Faced With Generator Use

Power Factor

A problem in the combination of a double-conversion UPS and a generator is the voltage distortion created by the UPS. The input of a double-conversion UPS is essentially a big rectifier. The current drawn by the UPS is non-sinusoidal. This can cause the voltage from the AC mains or a generator to also become non-sinusoidal. The voltage distortion then can cause problems in all electrical equipment connected to that power source, including the UPS itself. It will also cause more power to be lost in the wiring supplying power to the UPS due to the spikes in current flow. This level of "noise" is measured as a percentage of "total harmonic distortion of the current" (THD_I). Classic UPS rectifiers have a THD_I level of around 25%–30%. To reduce voltage distortion, this requires heavier mains wiring or generators more than twice as large as the UPS.

There are several solutions to reduce the THD_I in a double-conversion UPS:

Passive Power-Factor Correction

Classic solutions such as passive filters reduce THD_I to 5%–10% at full load. They are reliable, but big and only work at full load, and present their own problems when used in tandem with generators.

Active Power-Factor Correction

An alternative solution is an active filter. Through the use of such a device, THD_I can drop to 5% over the full power range. The newest technology in double-conversion UPS units is a rectifier that does not use classic rectifier components (thyristors and diodes) but uses high-frequency components instead. A double-conversion UPS with an insulated-gate bipolar transistor rectifier and inductor can have a THD_I as small as 2%. This completely eliminates the need to oversize the generator (and transformers), without additional filters, investment cost, losses, or space.

Communication

Power management (PM) requires

- The UPS to report its status to the computer it powers via a communications link such as a serial port, Ethernet and Simple Network Management Protocol, GSM/GPRS or USB
- A subsystem in the OS that processes the reports and generates notifications, PM events, or commands an ordered shut down. Some UPS manufacturers publish their communication protocols, but other manufacturers (such as APC) use proprietary protocols.

The basic computer-to-UPS control methods are intended for one-to-one signaling from a single source to a single target. For example, a single UPS may connect to a single computer to provide status information about the UPS, and allow the computer to control the UPS. Similarly, the USB protocol is also intended to connect a single computer to multiple peripheral devices.

In some situations it is useful for a single large UPS to be able to communicate with several protected devices. For traditional serial or USB control, a *signal replication* device may be used, which for example allows one UPS to connect to five computers using serial or USB connections. However, the splitting is typically only one direction from UPS to the devices to provide status information. Return control signals may only be permitted from one of the protected systems to the UPS.

As Ethernet has increased in common use since the 1990s, control signals are now commonly sent between a single UPS and multiple computers using standard Ethernet data communication methods such as TCP/IP. The status and control information is typically encrypted so that for example an outside hacker can not gain control of the UPS and command it to shut down.

Distribution of UPS status and control data requires that all intermediary devices such as Ethernet switches or serial multiplexers be powered by one or more UPS systems, in order for the UPS alerts to reach the target systems during a power outage. To avoid the dependency on Ethernet infrastructure, the UPSs can be connected directly to main control server by using GSM/GPRS channel also. The SMS or GPRS data packets sent from UPSs trigger software to shut down the PCs to reduce the load.

Batteries

Battery cabinet

The run-time for a battery-operated UPS depends on the type and size of batteries and rate of discharge, and the efficiency of the inverter. The total capacity of a lead–acid battery is a function of the rate at which it is discharged, which is described as Peukert's law.

Manufacturers supply run-time rating in minutes for packaged UPS systems. Larger systems (such as for data centers) require detailed calculation of the load, inverter efficiency, and battery characteristics to ensure the required endurance is attained.

Common Battery Characteristics and Load Testing

When a lead–acid battery is charged or discharged, this initially affects only the reacting chemicals, which are at the interface between the electrodes and the electrolyte. With time, the charge stored in the chemicals at the interface, often called "interface charge", spreads by diffusion of these chemicals throughout the volume of the active material.

If a battery has been completely discharged (e.g. the car lights were left on overnight) and next is given a fast charge for only a few minutes, then during the short charging time it develops only a charge near the interface. The battery voltage may rise to be close to the charger voltage so that the charging current decreases significantly. After a few hours this interface charge will spread to the volume of the electrode and electrolyte, leading to an interface charge so low that it may be insufficient to start a car.

Due to the interface charge, brief UPS *self-test* functions lasting only a few seconds may not accurately reflect the true runtime capacity of a UPS, and instead an extended *recalibration* or *rundown* test that deeply discharges the battery is needed.

The deep discharge testing is itself damaging to batteries due to the chemicals in the discharged battery starting to crystallize into highly stable molecular shapes that will not re-dissolve when the battery is recharged, permanently reducing charge capacity. In lead acid batteries this is known as sulfation but also affects other types such as nickel cadmium batteries and lithium batteries. Therefore, it is commonly recommended that rundown tests be performed infrequently, such as every six months to a year.

Testing of Strings of Batteries/Cells

Multi-kilowatt commercial UPS systems with large and easily accessible battery banks are capable of isolating and testing individual cells within a *battery string*, which consists of either combined-cell battery units (such as 12-V lead acid batteries) or individual chemical cells wired in series. Isolating a single cell and installing a jumper in place of it allows the one battery to be discharge-tested, while the rest of the battery string remains charged and available to provide protection.

It is also possible to measure the electrical characteristics of individual cells in a battery string, using intermediate sensor wires that are installed at every cell-to-cell junction, and monitored both individually and collectively. Battery strings may also be wired as series-parallel, for example two sets of 20 cells. In such a situation it is also necessary to monitor current flow between parallel strings, as current may circulate between the strings to balance out the effects of weak cells, dead cells with high resistance, or shorted cells. For example, stronger strings can discharge through weaker strings until voltage imbalances are equalized, and this must be factored into the individual inter-cell measurements within each string.

Series-Parallel Battery Interactions

Battery strings wired in series-parallel can develop unusual failure modes due to interactions between the multiple parallel strings. Defective batteries in one string can adversely affect the operation and lifespan of good or new batteries in other strings. These issues also apply to other situations where series-parallel strings are used, not just in UPS systems but also in electric vehicle applications.

Consider a series-parallel battery arrangement with all good cells, and one becomes shorted or dead:

- The failed cell will reduce the maximum developed voltage for the entire series string it is within.
- Other series strings wired in parallel with the degraded string will now discharge through the degraded string until their voltage matches the voltage of the degraded string, potentially overcharging and leading to electrolyte boiling and outgassing from the remaining good cells in the degraded string. These parallel strings can now never be fully recharged, as the increased voltage will bleed off through the string containing the failed battery.
- Charging systems may attempt to gauge battery string capacity by measuring overall voltage. Due to the overall string voltage depletion due to the dead cells, the charging system may detect this as a state of discharge, and will continuously attempt to charge the series-parallel strings, which leads to continuous overcharging and damage to all the cells in the degraded series string containing the damaged battery.
- If lead-acid batteries are used, all cells in the formerly good parallel strings will begin to sulfate due to the inability for them to be fully recharged, resulting in the storage capacity of these cells being permanently damaged, even if the damaged cell in the one degraded string is eventually discovered and replaced with a new one.

The only way to prevent these subtle series-parallel string interactions is by not using parallel strings at all and using separate charge controllers and inverters for individual series strings.

Series New/Old Battery Interactions

Even just a single string of batteries wired in series can have adverse interactions if new batteries are mixed with old batteries. Older batteries tend to have reduced storage capacity, and so will both discharge faster than new batteries and also charge to their maximum capacity more rapidly than new batteries.

As a mixed string of new and old batteries is depleted, the string voltage will drop, and when the old batteries are exhausted the new batteries still have charge available. The newer cells may continue to discharge through the rest of the string, but due to the low voltage this energy flow may not be useful, and may be wasted in the old cells as resistance heating.

For cells that are supposed to operate within a specific discharge window, new cells with more capacity may cause the old cells in the series string to continue to discharge beyond the safe bottom limit of the discharge window, damaging the old cells.

When recharged, the old cells recharge more rapidly, leading to a rapid rise of voltage to near the fully charged state, but before the new cells with more capacity have fully recharged. The charge controller detects the high voltage of a nearly fully charged string and reduces current flow. The new cells with more capacity now charge very slowly, so slowly that the chemicals may begin to crystallize before reaching the fully charged state, reducing new cell capacity over several charge/discharge cycles until their capacity more closely matches the old cells in the series string.

For such reasons, some industrial UPS management systems recommend periodic replacement of entire battery arrays potentially using hundreds of expensive batteries, due to these damaging interactions between new batteries and old batteries, within and across series and parallel strings.

SOLAR INVERTER

A solar inverter, or converter or PV inverter, converts the variable direct current (DC) output of a photovoltaic (PV) solar panel into a utility frequency alternating current (AC) that can be fed into a commercial electrical grid or used by a local, off-grid electrical network. It is a critical balance of system (BOS)–component in a photovoltaic system, allowing the use of ordinary AC-powered equipment. Solar power inverters have special functions adapted for use with photovoltaic arrays, including maximum power point tracking and anti-islanding protection.

Internal view of a solar inverter. Note the many large capacitors (blue cylinders), used to store energy briefly and improve the output waveform.

Classification

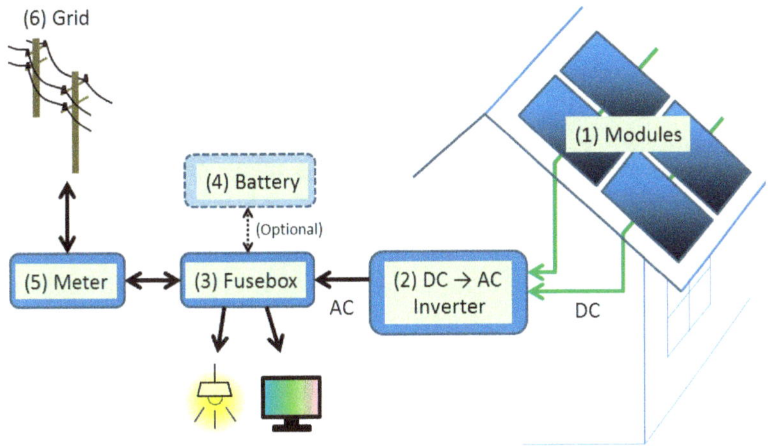

Simplified schematics of a grid-connected residential photovoltaic power system.

Solar inverters may be classified into three broad types:

- Stand-alone inverters, used in isolated systems where the inverter draws its DC energy from batteries charged by photovoltaic arrays. Many stand-alone inverters also incorporate integral battery chargers to replenish the battery from an AC source, when available. Normally these do not interface in any way with the utility grid, and as such, are not required to have anti-islanding protection.
- Grid-tie inverters, which match phase with a utility-supplied sine wave. Grid-tie inverters are designed to shut down automatically upon loss of utility supply, for safety reasons. They do not provide backup power during utility outages.
- Battery backup inverters, are special inverters which are designed to draw energy from a battery, manage the battery charge via an onboard charger, and export excess energy to the utility grid. These inverters are capable of supplying AC energy to selected loads during a utility outage, and are required to have anti-islanding protection.

Maximum Power Point Tracking

Maximum power point tracking (MPPT or sometimes just PPT) is a technique used commonly with wind turbines and photovoltaic (PV) solar systems to maximize power extraction under all conditions.

Although solar power is mainly covered, the principle applies generally to sources with variable power: for example, optical power transmission and thermophotovoltaics.

PV solar systems exist in many different configurations with regard to their relationship to inverter systems, external grids, battery banks, or other electrical loads. Regardless of the ultimate destination of the solar power, though, the central problem addressed by MPPT is that the efficiency of power transfer from the solar cell depends on both the amount of sunlight falling on the solar panels and the electrical characteristics of the load. As the amount of sunlight varies, the load characteristic that gives the highest power transfer efficiency changes, so that the efficiency of the system is optimized when the load characteristic changes to keep the power transfer at highest efficiency. This load characteristic is called the **maximum power point** and MPPT is the process of finding this point and keeping the load characteristic there. Electrical circuits can be designed to present arbitrary loads to the photovoltaic cells and then convert the voltage, current, or frequency to suit other devices or systems, and MPPT solves the problem of choosing the best load to be presented to the cells in order to get the most usable power out.

Solar cells have a complex relationship between temperature and total resistance that produces a non-linear output efficiency which can be analyzed based on the I-V curve. It is the purpose of the MPPT system to sample the output of the PV cells and apply the proper resistance (load) to obtain maximum power for any given environmental conditions. MPPT devices are typically integrated into an electric power converter system that provides voltage or current conversion, filtering, and regulation for driving various loads, including power grids, batteries, or motors.

- Solar inverters convert the DC power to AC power and may incorporate MPPT: such inverters sample the output power (I-V curve) from the solar modules and apply the proper resistance (load) so as to obtain maximum power.
- The power at the MPP (P_{mpp}) is the product of the MPP voltage(V_{mpp}) and MPP current(I_{mpp}).

I-V Curve

Photovoltaic cells have a complex relationship between their operating environment and the maximum power they can produce. The fill factor, abbreviated *FF*, is a parameter which characterizes the non-linear electrical behavior of the solar cell. Fill factor is defined as the ratio of the maximum power from the solar cell to the product of Open Circuit Voltage V_{oc} and Short-Circuit Current I_{sc}. In tabulated data it is often used to estimate the maximum power that a cell can provide with an optimal load under given conditions, $P=FF*V_{oc}*I_{sc}$. For most purposes, FF, V_{oc}, and I_{sc} are enough information to give a useful approximate model of the electrical behavior of a photovoltaic cell under typical conditions.

For any given set of operational conditions, cells have a single operating point where the values of the current (*I*) and Voltage (*V*) of the cell result in a maximum power output. These values correspond to a particular load resistance, which is equal to *V* / I as specified by Ohm's Law. The power P is given by $P=V*I$. A photovoltaic cell, for the majority of its useful curve, acts as a constant current source. However, at a photovoltaic cell's MPP region, its curve has an approximately inverse exponential relationship between current and voltage. From basic circuit theory, the power delivered from or to a device is optimized where the derivative (graphically, the slope) dI/dV of the I-V curve is equal and opposite the I/V ratio (where $dP/dV=0$). This is known as the maximum power point (MPP) and corresponds to the "knee" of the curve.

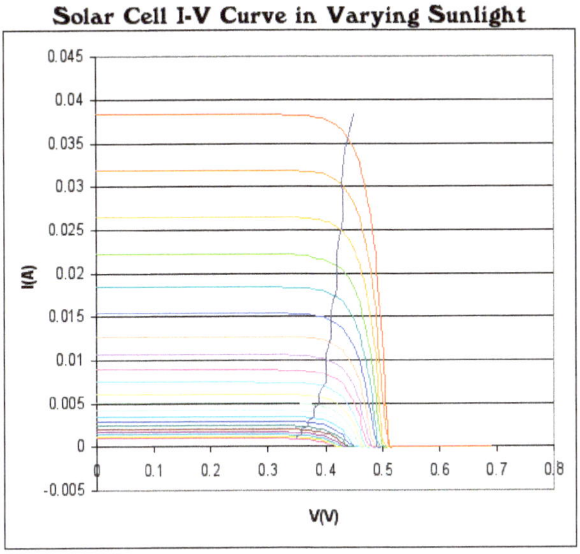

Photovoltaic solar cell I-V curves where a line intersects the knee of the curves where the maximum power transfer point is located.

A load with resistance $R=V/I$ equal to the reciprocal of this value draws the maximum power from the device. This is sometimes called the 'characteristic resistance' of the cell. This is a dynamic quantity which changes depending on the level of illumination, as well as other factors such as temperature and the age of the cell. If the resistance is lower or higher than this value, the power drawn will be less than the maximum available, and thus the cell will not be used as efficiently as it could be. Maximum power point trackers utilize different types of control circuit or logic to search for this point and thus to allow the converter circuit to extract the maximum power available from a cell.

MPPT Implementation

When a load is directly connected to the solar panel, the operating point of the panel will rarely be at peak power. The impedance seen by the panel derives the operating point of the solar panel. Thus by varying the impedance seen by the panel, the operating point can be moved towards peak power point. Since panels are DC devices, DC-DC converters must be utilized to transform the impedance of one circuit (source) to the other circuit (load). Changing the duty ratio of the DC-DC converter results in an impedance change as seen by the panel. At a particular impedance (or duty ratio) the operating point will be at the peak power transfer point. The I-V curve of the panel can vary considerably with variation in atmospheric conditions such as radiance and temperature. Therefore it is not feasible to fix the duty ratio with such dynamically changing operating conditions.

MPPT implementations utilize algorithms that frequently sample panel voltages and currents, then adjust the duty ratio as needed. Microcontrollers are employed to implement the algorithms. Modern implementations often utilize larger computers for analytics and load forcasting.

Classification

Controllers can follow several strategies to optimize the power output of an array. Maximum power point trackers may implement different algorithms and switch between them based on the operating conditions of the array.

Perturb and Observe

In this method the controller adjusts the voltage by a small amount from the array and measures power; if the power increases, further adjustments in that direction are tried until power no longer increases. This is called the perturb and observe method and is most common, although this method can result in oscillations of power output. It is referred to as a *hill climbing* method, because it depends on the rise of the curve of power against voltage below the maximum power point, and the fall above that point. Perturb and observe is the most commonly used MPPT method due to its ease of implementation. Perturb and observe method may result in top-level efficiency, provided that a proper predictive and adaptive hill climbing strategy is adopted.

Incremental Conductance

In the incremental conductance method, the controller measures incremental changes in PV array current and voltage to predict the effect of a voltage change. This method requires more computation in the controller, but can track changing conditions more rapidly than the perturb and observe method (P&O). Like the P&O algorithm, it can produce oscillations in power output. This method utilizes the incremental conductance (dI/dV) of the photovoltaic array to compute the sign of the change in power with respect to voltage (dP/dV).

The incremental conductance method computes the maximum power point by comparison of the incremental conductance (I_Δ / V_Δ) to the array conductance (I / V). When these two are the same (I / V = I_Δ / V_Δ), the output voltage is the MPP voltage. The controller maintains this voltage until the irradiation changes and the process is repeated.

Current Sweep

The current sweep method uses a sweep waveform for the PV array current such that the I-V characteristic of the PV array is obtained and updated at fixed time intervals. The maximum power point voltage can then be computed from the characteristic curve at the same intervals.

Constant Voltage

The term "constant voltage" in MPP tracking is used to describe different techniques by different authors, one in which the output voltage is regulated to a constant value under all conditions and one in which the output voltage is regulated based on a constant ratio to the measured open circuit voltage (V_{OC}). The latter technique is referred to in contrast as the "open voltage" method by some authors. If the output voltage is held constant, there is no attempt to track the maximum power point, so it is not a maximum power point tracking technique in a strict sense, though it does have some advantages in cases when the MPP tracking tends to fail, and thus it is sometimes used to supplement an MPPT method in those cases.

In the "constant voltage" MPPT method (also known as the "open voltage method"), the power delivered to the load is momentarily interrupted and the open-circuit voltage with zero current is measured. The controller then resumes operation with the voltage controlled at a fixed ratio, such as 0.76, of the open-circuit voltage V_{OC}. This is usually a value which has been determined to be the maximum power point, either empirically or based on modelling, for expected operating conditions. The operating point of the PV array is thus kept near the MPP by regulating the array voltage and matching it to the fixed reference voltage $V_{ref}=kV_{OC}$. The value of V_{ref} may be also chosen to give optimal performance relative to other factors as well as the MPP, but the central idea in this technique is that V_{ref} is determined as a ratio to V_{OC}.

One of the inherent approximations to the "constant voltage" ratio method is that the ratio of the MPP voltage to V_{OC} is only approximately constant, so it leaves room for further possible optimization.

Comparison of Methods

Both perturb and observe, and incremental conductance, are examples of "hill climbing" methods that can find the local maximum of the power curve for the operating condition of the PV array, and so provide a true maximum power point.

The perturb and observe method requires oscillating power output around the maximum power point even under steady state irradiance.

The incremental conductance method has the advantage over the perturb and observe (P&O) method that it can determine the maximum power point without oscillating around this value. It can perform maximum power point tracking under rapidly varying irradiation conditions with higher accuracy than the perturb and observe method. However, the incremental conductance method can produce oscillations (unintentionally) and can perform erratically under rapidly changing atmospheric conditions. The sampling frequency is decreased due to the higher complexity of the algorithm compared to the P&O method.

In the constant voltage ratio (or "open voltage") method, the current from the photovoltaic array must be set to zero momentarily to measure the open circuit voltage and then afterwards set to a predetermined percentage of the measured voltage, usually around 76%. Energy may be wasted during the time the current is set to zero. The approximation of 76% as the MPP/V_{OC} ratio is not necessarily accurate though. Although simple and low-cost to implement, the interruptions reduce array efficiency and do not ensure finding the actual maximum power point. However, efficiencies of some systems may reach above 95%.

MPPT Placement

Traditional solar inverters perform MPPT for the entire PV array (module association) as a whole. In such systems the same current, dictated by the inverter, flows through all modules in the string (series). Because different modules have different I-V curves and different MPPs (due to manufacturing tolerance, partial shading, etc.) this architecture means some modules will be performing below their MPP, resulting in lower efficiency.

Some companies (see power optimizer) are now placing maximum power point tracker into individual modules, allowing each to operate at peak efficiency despite uneven shading, soiling or electrical mismatch.

Data suggests having one inverter with one MPPT for a project that has east and west-facing modules presents no disadvantages when compared to having two inverters or one inverter with more than one MPPT.

Operation with Batteries

At night, an off-grid PV system may use batteries to supply loads. Although the fully charged battery pack voltage may be close to the PV panel's maximum power point voltage, this is unlikely to be true at sunrise when the battery has been partially discharged. Charging may begin at a voltage considerably below the PV panel maximum power point voltage, and an MPPT can resolve this mismatch.

When the batteries in an off-grid system are fully charged and PV production exceeds local loads, an MPPT can no longer operate the panel at its maximum power point as the excess power has no load to absorb it. The MPPT must then shift the PV panel operating point away from the peak power point until production exactly matches demand. (An alternative approach commonly used in spacecraft is to divert surplus PV power into a resistive load, allowing the panel to operate continuously at its peak power point.)

In a grid connected photovoltaic system, all delivered power from solar modules will be sent to the grid. Therefore, the MPPT in a grid connected PV system will always attempt to operate the PV modules at its maximum power point.

Solar Micro-Inverters

A solar micro-inverter, or simply microinverter, is a device used in photovoltaics that converts direct current (DC) generated by a single solar module to alternating current (AC). The output from several microinverters is combined and often fed to the electrical grid. Microinverters contrast with conventional string and central solar inverters, which are connected to multiple solar modules or panels of the PV system.

Microinverters have several advantages over conventional inverters. The main advantage is that small amounts of shading, debris or snow lines on any one solar module, or even a complete module failure, do not disproportionately reduce the output of the entire array. Each microinverter harvests optimum power by performing maximum power point tracking for its connected module. Simplicity in system design, lower amperage wires, simplified stock management, and added safety are other factors introduced with the microinverter solution.

The primary disadvantages of a microinverter include a higher initial equipment cost per peak watt than the equivalent power of a central inverter since each inverter needs to be installed adjacent to a panel (usually on a roof). This also makes them harder to maintain and more costly to remove and replace (O&M). Some manufacturers have addressed these issues with panels with built-in microinverters.

A type of technology similar to a microinverter is a power optimizer which also does panel-level maximum power point tracking, but does not convert to AC per module.

A solar micro-inverter in the process of being installed. The ground wire is attached to the lug and the panel's DC connections are attached to the cables on the lower right. The AC parallel trunk cable runs at the top (just visible).

Description

Solar Inverter

Solar panels produce direct current at a voltage that depends on module design and lighting conditions. Modern modules using 6-inch cells typically contain 60 cells and produce a nominal 30 V. For conversion into AC, panels may be connected in series to produce an array that is effectively a single large panel with a nominal rating of 300 to 600 VDC. The power then runs to an inverter, which converts it into standard AC voltage, typically 230 VAC / 50 Hz or 240 VAC / 60 Hz.

The main problem with the "string inverter" approach is the string of panels acts as if it were a single larger panel with a max current rating equivalent to the poorest performer in the string. For example, if one panel in a string has 5% higher resistance due to a minor manufacturing defect, the entire string suffers a 5% performance loss. This situation is dynamic. If a panel is shaded its output drops dramatically, affecting the output of the string, even if the other panels are not shaded. Even slight changes in orientation can cause output loss in this fashion. In the industry, this is known as the "Christmas-lights effect", referring to the way an entire string of series-strung Christmas tree lights will fail if a single bulb fails. However, this effect is not entirely accurate and ignores the complex interaction between modern string inverter maximum power point tracking and even module bypass diodes. Shade studies by major micro inverter and DC optimizer companies show small yearly gains in light, medium and heavy shaded conditions- 2%, 5% and 8% respectively- over an older string inverter.

Additionally, the efficiency of a panel's output is strongly affected by the load the inverter places on it. To maximize production, inverters use a technique called maximum power point tracking (MPPT) to ensure optimal energy harvest by adjusting the applied load. However, the same issues that cause output to vary from panel to panel, affect the proper load that the MPPT system should apply. If a single panel operates at a different point, a string inverter can only see the overall change, and moves the MPPT point to match. This results in not just losses from the shadowed panel, but the other panels too. Shading of as little as 9% of the surface of an array can, in some circumstances, reduce system-wide power as much as 54%. However, as stated above, these yearly yield losses are relatively small and newer technologies allow some string inverters to significantly reduce the effects of partial shading.

Another issue, though minor, is that string inverters are available in a limited selection of power ratings. This means that a given array normally up-sizes the inverter to the next-largest model over the rating of the panel array. For instance, a 10-panel array of 2300 W might have to use a 2500 or even 3000 W inverter, paying for conversion capability it cannot use. This same issue makes it difficult to change array size over time, adding power when funds are available. If the customer originally purchased a 2500 W inverter for their 2300 W of panels, they cannot add even a single panel without over-driving the inverter. However, this over sizing is considered common practice in today's industry (sometimes as high as 20% over inverter nameplate rating) to account for module degradation, higher performance during winter months or to achieve higher sell back to the utility.

Other challenges associated with centralized inverters include the space required to locate the device, as well as heat dissipation requirements. Large central inverters are typically actively cooled. Cooling fans make noise, so location of the inverter relative to offices and occupied areas must be considered. And because cooling fans have moving parts, dirt, dust, and moisture can negatively affect their performance over time. String inverters are quieter but might produce a humming noise in late afternoon when inverter power is low.

Microinverter Concept

Microinverters are small inverters rated to handle the output of a single panel. Modern grid-tie panels are normally rated between 225 and 275W, but rarely produce this in practice, so microinverters are typically rated between 190 and 220 W. Because it is operated at this lower power point, many design issues inherent to larger designs simply go away; the need for a large transformer is generally eliminated, large electrolytic capacitors can be replaced by more reliable thin-film capacitors, and cooling loads are reduced so no fans are needed. Mean time between failures (MTBF) are quoted in hundreds of years.

More importantly, a microinverter attached to a single panel allows it to isolate and tune the output of that panel. For example, in the same 10-panel array used as an example above, with microinverters any panel that is under-performing has no effect on panels around it. In that case, the array as a whole produces as much as 5% more power than it would with a string inverter. When shadowing is factored in, if present, these gains can become considerable, with manufacturers generally claiming 5% better output at a minimum, and up to 25% better in some cases. Furthermore,a single model can be used with a wide variety of panels, new panels can be added to an array at any time, and do not have to have the same rating as existing panels.

Microinverters produce grid-matching power directly at the back of the panel. Arrays of panels are connected in parallel to each other, and then to the grid. This has the major advantage that a single failing panel or inverter cannot take the entire string offline. Combined with the lower power and heat loads, and improved MTBF, some suggest that overall array reliability of a microinverter-based system is significantly greater than a string inverter-based one. This assertion is supported by longer warranties, typically 15 to 25 years, compared with 5 or 10 year warranties that are more typical for string inverters. Additionally, when faults occur, they are identifiable to a single point, as opposed to an entire string. This not only makes fault isolation easier, but unmasks minor problems that might not otherwise become visible – a single under-performing panel may not affect a long string's output enough to be noticed.

Disadvantages

The main disadvantage of the microinverter concept has, until recently, been cost. Because each microinverter has to duplicate much of the complexity of a string inverter but spread that out over a smaller power rating, costs on a per-watt basis are greater. This offsets any advantage in terms of simplification of individual components. As of October 2010, a central inverter costs approximately $0.40 per watt, whereas a microinverter costs approximately $0.52 per watt. Like string inverters, economic considerations force manufacturers to limit the number of models they produce. Most produce a single model that may be over or under-size when matched with a specific panel.

In many cases the packaging can have a significant effect on price. With a central inverter you may have only one set of panel connections for dozens of panels, a single AC output, and one box. With microinverters, each one has to have its own set of inputs and outputs, in its own box. Because that box is on the roof, it has to be sealed and weatherproofed. This can represent a significant portion of the overall price-per-watt.

To further reduce costs, some models control two or three panels from a single box, reducing the packaging and associated costs. Some systems simply place two entire micros in a single box, while others duplicate only the MPPT section of the system and use a single DC-to-AC stage for further cost reductions. Some have suggested that this approach will make microinverters comparable in cost with those using string inverters. With steadily decreasing prices, the introduction of dual microinverters and the advent of wider model selections to match PV module output more closely, cost is less of an obstacle so microinverters may now spread more widely.

Microinverters have become common where array sizes are small and maximizing performance from every panel is a concern. In these cases, differential in price-per-watt is minimized due to the small number of panels, and has little effect on overall system cost.

The improvement in energy harvest given a fixed size array can offset this difference in cost. For this reason, microinverters have been most successful in the residential market, where limited space for panels constrains array size, and shading from nearby trees or other objects is often an issue. Microinverter manufacturers list many installations, some as small as a single panel and the majority under 50.

An often overlooked disadvantage of micro inverters is the future O&M costs associated with them. While the technology has improved over the years the fact remains that the devices will eventually either fail or wear out. The installer must balance these replacement costs (around $400 per truck roll), increased safety risks to personnel, equipment and module racking against the profit margins for the installation. For homeowners, the eventual wear out or premature device failures will introduce potential damage to the roof tiles or shingles, property damage and other nuisances.

Advantages

While microinverters generally have a lower efficiency than string inverters, the overall efficiency is increased due to the fact that every inverter / panel unit acts independently. In a string configuration, when an panel on a string is shaded, the output of the entire string of panels is reduced to the output of the lowest producing panel. This is not the case with micro inverters.

A further advantage is found in the panel output quality. The rated output of any two panels in the same production run can vary by as much as 10% or more. This is mitigated with a string configuration but not so in a microinverter configuration. The result is maximum power harvesting from a microinverter array.

Monitoring and maintenance is also easier as many microinverter producers provide apps or websites to monitor the power output of their units. In many cases, these are proprietary; however this is not always the case. Following the demise of Enecsys, and the subsequent closure of their site; a number of private sites such as Enecsys-Monitoring sprung up to enable owners to continue to monitor their systems.

Three-Phase Microinverters

Efficient conversion of DC power to AC requires the inverter to store energy from the panel while the grid's AC voltage is near zero, and then release it again when it rises. This requires considerable amounts of energy storage in a small package. The lowest-cost option for the required amount of storage is the electrolytic capacitor, but these have relatively short lifetimes normally measured in years, and those lifetimes are shorter when operated hot, like on a rooftop solar panel. This has led to considerable development effort on the part of microinverter developers, who have introduced a variety of conversion topologies with lowered storage requirements, some using the much less capable but far longer lived film capacitors where possible.

Three-phase electric power represents another solution to the problem. In a three-phase circuit, the power does not vary between (say) +120 to -120 Volts between two lines, but instead varies between 60 and +120 or -60 and -120V, and the periods of variation are much shorter. Inverters designed to operate on three phase systems require much less storage. A three-phase micro using zero-voltage switching can also offer higher circuit density and lower cost components, while improving conversion efficiency to over 98%, better than the typical one-phase peak around 96%.

Three-phase systems, however, are generally only seen in industrial and commercial settings. These markets normally install larger arrays, where price sensitivity is the highest. Uptake of three-phase micros, in spite of any theoretical advantages, appears to be very low.

History

The microinverter concept has been in the solar industry since its inception. However, flat costs in manufacturing, like the cost of the transformer or enclosure, scaled favorably with size, and meant that larger devices were inherently less expensive in terms of price per watt. Small inverters were available from companies like ExelTech and others, but these were simply small versions of larger designs with poor price performance, and were aimed at niche markets.

Early Examples

In 1991 the US company Ascension Technology started work on what was essentially a shrunken version of a traditional inverter, intended to be mounted on a panel to form an *AC panel*. This design was based on the conventional linear regulator, which is not particularly efficient and dissipates considerable heat. In 1994 they sent an example to Sandia Labs for testing. In 1997, Ascension partnered with US panel company ASE Americas to introduce the 300 W SunSine panel.

Released in 1993, Mastervolt's Sunmaster 130S was the first true microinverter.

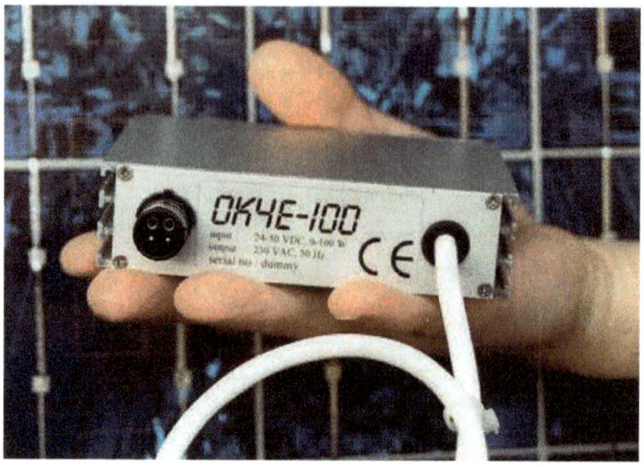

Another early microinverter, 1995's OK4E-100 – E for European, 100 for 100 watts.

Design of, what would today be recognized as a "true" microinverter, traces its history to late 1980s work by Werner Kleinkauf at the ISET (*Institut für Solare Energieversorgungstechnik*), now Fraunhofer Institute for Wind Energy and Energy System Technology. These designs were based on modern high-frequency switching power supply technology, which is much more efficient. His work on "module integrated converters" was highly influential, especially in Europe.

In 1993 Mastervolt introduced their first grid-tie inverter, the Sunmaster 130S, based on a collaborative effort between Shell Solar, Ecofys and ECN. The 130 was designed to mount directly to the back of the panel, connecting both AC and DC lines with compression fittings. In 2000, the 130 was replaced by the Soladin 120, a microinverter in the form of an AC adapter that allows panels to be connected simply by plugging them into any wall socket.

In 1995, OKE-Services designed a new high-frequency version with improved efficiency, which was introduced commercially as the OK4-100 in 1995 by NKF Kabel, and re-branded for US sales as the Trace Microsine. A new version, the OK4All, improved efficiency and had wider operating ranges.

In spite of this promising start, by 2003 most of these projects had ended. Ascension Technology was purchased by Applied Power Corporation, a large integrator. APC was in turn purchased by Schott in 2002, and SunSine production was canceled in favor of Schott's existing designs. NKF ended production of the OK4 series in 2003 when a subsidy program ended. Mastervolt has moved on to a line of "mini-inverters" combining the ease-of-use of the 120 in a system designed to support up to 600 W of panels.

Enphase

In the aftermath of the 2001 Telecoms crash, Martin Fornage of Cerent Corporation was looking for new projects. When he saw the low performance of the string inverter for the solar array on his ranch, he found the project he was looking for. In 2006 he formed Enphase Energy with another Cerent engineer, Raghu Belur, and they spent the next year applying their telecommunications design expertise to the inverter problem.

Released in 2008, the Enphase M175 model was the first commercially successful microinverter. A successor, the M190, was introduced in 2009, and the latest model, the M215, in 2011. Backed by $100 million in private equity, Enphase quickly grew to 13% marketshare by mid-2010, aiming for 20% by year-end. They shipped their 500,000th inverter in early 2011,and their 1,000,000th in September of the same year. In early 2011, they announced that re-branded versions of the new design will be sold by Siemens directly to electrical contractors for widespread distribution.

Enphase has subscribed an agreement with EnergyAustralia, to market its micro-inverter technology.

Competition

Enphase's success did not go unnoticed, and since 2010 a host of competitors have appeared. Many of these are identical to the M190 in specs, and even in the casing and mounting details. Some differentiate by competing head-to-head with Enphase in terms of price or performance, while others are attacking niche markets.

Larger firms have also stepped into the field; OKE-Services updated OK4-All product was recently bought by SMA and released as the SunnyBoy 240 after an extended gestation period, while Power-One has introduced the AURORA 250 and 300. Other major players included Enecsys and SolarBridge, especially outside the North American market. The only USA made microinverter in production is from Chilicon Power. Since 2009, several companies from Europe to China, including major central inverter manufacturers, have launched microinverters—validating the microinverter as an established technology and one of the biggest technology shifts in the PV industry in recent years.

Price Issues

The period between 2009 and 2012 included unprecedented downward price movement in the PV market. At the beginning of this period, panels were generally around $2.00 to $2.50/W, and inverters around 50 to 65 cents/W. By the end of 2012, panels were widely available in wholesale at 65 to 70 cents, and string inverters around 30 to 35 cents/W. In comparison, microinverters have proven relatively immune to these same sorts of price declines, moving from about 65 cents/W to 50 to 55 once cabling is factored in. This has led to widening losses as the suppliers attempt to remain competitive.

Grid Tied Solar Inverters

A grid-tie inverter converts direct current (DC) electricity into an alternating current (AC) suitable for injecting into an electrical power grid, normally 120V RMS at 60Hz or 240V RMS at 50Hz. Grid-tie inverters are used between local electrical power generators: solar panel, wind turbine, hydro-electric, and the grid.

In order to inject electrical power efficiently and safely into the grid, grid-tie inverters must accurately match the voltage and phase of the grid sine wave AC waveform.

Some electricity companies will pay for electrical power that is injected into the grid.

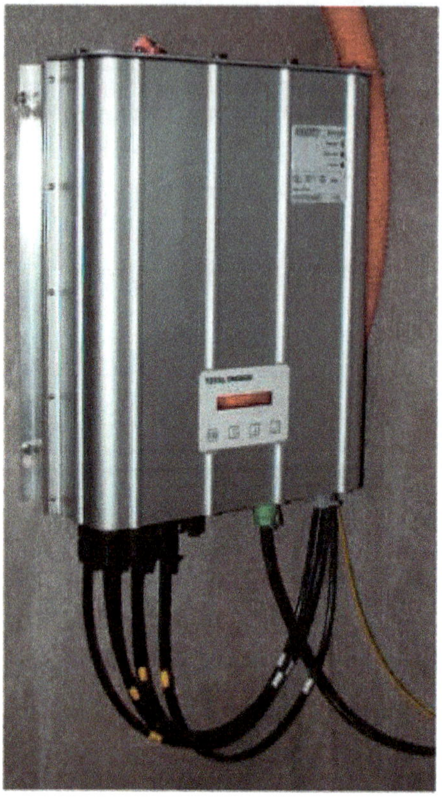

Inverter for grid-tied solar panel.

Payment for Injected Power

Electricity companies, in some countries, pay for electrical power that is injected into the electricity utility grid. Payment is arranged in several ways.

With net metering the electricity company pays for the net power injected into the grid, as recorded by a meter in the customer's premises. For example, a customer may consume 400 kilowatt-hours over a month and may return 500 kilowatt-hours to the grid in the same month. In this case the electricity company would pay for the 100 kilowatt hours balance of power fed back into the grid. In the US, net metering policies vary by jurisdiction.

Feed-in tariff, based on a contract with a distribution company or other power authority, is where the customer is paid for electrical power injected into the grid.

In the United States, grid-interactive power systems are specified in the National Electric Code, which also mandates requirements for grid-interactive inverters.

Three-phase grid-tie inverter for large solar panel systems.

Operation

Grid-tie inverters convert DC electrical power into AC power suitable for injecting into the electric utility company grid. The grid tie inverter (GTI) must match the phase of the grid and maintain the output voltage slightly higher than the grid voltage at any instant. A high-quality modern grid-tie inverter has a fixed unity power factor, which means its output voltage and current are perfectly lined up, and its phase angle is within 1 degree of the AC power grid. The inverter has an on-board computer which senses the current AC grid waveform, and outputs a voltage to correspond with the grid. However, supplying reactive power to the grid might be necessary to keep the voltage in the local grid inside allowed limitations. Otherwise, in a grid segment with considerable power from renewable sources, voltage levels might rise too much at times of high production, i.e. around noon with solar panels.

Grid-tie inverters are also designed to quickly disconnect from the grid if the utility grid goes down. This is an NEC requirement that ensures that in the event of a blackout, the grid tie inverter will shut down to prevent the energy it transfers from harming any line workers who are sent to fix the power grid.

Properly configured, a grid tie inverter enables a home owner to use an alternative power generation system like solar or wind power without extensive rewiring and without batteries. If the alternative power being produced is insufficient, the deficit will be sourced from the electricity grid.

Types

Grid-tie inverters include conventional low-frequency types with transformer coupling, newer high-frequency types, also with transformer coupling, and transformerless types. Instead of converting direct current directly into AC suitable for the grid, high-frequency transformers types use a computer process to convert the power to a high-frequency and then back to DC and then to the final AC output voltage suitable for the grid.

Inside an SWEA 250W transformer-coupled grid-tie inverter.

Transformerless inverters, which are popular in Europe, are lighter, smaller, and more efficient than inverters with transformers. But transformerless inverters have been slow to enter the US market because of concerns that transformerless inverters, which do not have galvanic isolation between the DC side and grid, could inject dangerous DC voltages and currents into the grid under fault conditions.

However, since 2005, the NFPA's NEC allows transformerless, or non-galvanically isolated, inverters by removing the requirement that all solar electric systems be negative grounded and specifying new safety requirements. Amendments to VDE 0126-1-1 and IEC 6210 define the design and procedures needed for such systems: primarily, ground current measurement and DC to grid isolation tests.

Datasheets

Manufacturers datasheets for their inverters usually include the following data:

- *Rated output power*: This value is provided in watts or kilowatts. For some inverters, they may provide an output rating for different output voltages. For instance, if the inverter can be configured for either 240 VAC or 208 VAC output, the rated power output may be different for each of those configurations.
- *Output voltage(s)*: This value indicates to which utility voltages the inverter can

connect. For smaller inverters that are designed for residential use, the output voltage is usually 240 VAC. Inverters that target commercial applications are rated for 208, 240, 277, 400, 480 or 600 VAC and may also produce three phase power.

- *Peak efficiency*: The peak efficiency represents the highest efficiency that the inverter can achieve. Most grid-tie inverters on the market as of July 2009 have peak efficiencies of over 94%, some as high as 96%. The energy lost during inversion is for the most part converted into heat. Consequently, in order for an inverter to output its rated power it will require a power input that exceeds its output. For example, a 5000 W inverter operating at full power at 95% efficiency will require an input of 5,263 W (rated power divided by efficiency). Inverters that are capable of producing power at different AC voltages may have different efficiencies associated with each voltage.
- *CEC weighted efficiency*: This efficiency is published by the California Energy Commission on its GoSolar website. In contrast to peak efficiency, this value is an average efficiency and is a better representation of the inverter's operating profile. Inverters that are capable of producing power at different AC voltages may have different efficiencies associated with each voltage.
- *Maximum input current*: This is the maximum amount of direct current that the inverter can use. If a system, solar cells for example, produces a current in excess of the maximum input current, that current is not used by the inverter.
- *Maximum output current*: The maximum output current is the maximum continuous alternating current that the inverter will supply. This value is typically used to determine the minimum current rating of the over-current protection devices (e.g., breakers and fuses) and disconnects required for the output circuit. Inverters that are capable of producing power at different AC voltages will have different maximum outputs for each voltage.
- *Peak power tracking voltage*: This represents the DC voltage range in which the inverter's maximum point power tracker will operate. The system designer must configure the strings optimally so that during the majority of the year, the voltage of the strings will be within this range. This can be a difficult task since voltage will fluctuate with changes in temperature.
- *Start voltage*: This value is not listed on all inverter datasheets. The value indicates the minimum DC voltage that is required in order for the inverter to turn on and begin operation. This is especially important for solar applications, because the system designer must be sure that there is a sufficient number of solar modules wired in series in each string to produce this voltage. If this value is not provided by the manufacturer, system designers typically use the lower band of the peak power tracking voltage range as the inverter's minimum voltage.
- *IPxx rating*: The Ingress Protection rating or IP Code classifies and rates the level of protection provided against the ingress of solid foreign objects (first digit) or water (second digit), a higher digit means greater protection. In the US the *NEMA enclosure type* is used similarly to the international rating. Most inverters are rated for outdoors installation with IP45 (no dust protection) or IP65

(dust tight), or in the US, NEMA 3R (no windblown dust protection) or NEMA 4X (windblown dust, direct water splash and additional corrosion protection).

Solar Charge Controller

A charge controller, charge regulator or battery regulator limits the rate at which electric current is added to or drawn from electric batteries. It prevents overcharging and may protect against overvoltage, which can reduce battery performance or lifespan, and may pose a safety risk. It may also prevent completely draining ("deep discharging") a battery, or perform controlled discharges, depending on the battery technology, to protect battery life. The terms "charge controller" or "charge regulator" may refer to either a stand-alone device, or to control circuitry integrated within a battery pack, battery-powered device, or battery charger.

A solar charge controller.

Stand-Alone Charge Controllers

Charge controllers are sold to consumers as separate devices, often in conjunction with solar or wind power generators, for uses such as RV, boat, and off-the-grid home battery storage systems. In solar applications, charge controllers may also be called solar regulators. Some charge controllers / solar regulators have additional features, such as a low voltage disconnect (LVD), a separate circuit which powers down the load when the batteries become overly discharged (some battery chemistries are such that over-discharge can ruin the battery).

A series charge controller or series regulator disables further current flow into batteries when they are full. A shunt charge controller or shunt regulator diverts excess electricity to an auxiliary or "shunt" load, such as an electric water heater, when batteries are full.

Simple charge controllers stop charging a battery when they exceed a set high voltage level, and re-enable charging when battery voltage drops back below that level. Pulse width modulation (PWM) and maximum power point tracker (MPPT) technologies are more electronically sophisticated, adjusting charging rates depending on the battery's level, to allow charging closer to its maximum capacity.

A charge controller with MPPT capability frees the system designer from closely matching available PV voltage to battery voltage. Considerable efficiency gains can be achieved, particularly when the PV array is located at some distance from the battery. By way of example, a 150 volt PV array connected to an MPPT charge controller can be used to charge a 24 or 48 volt battery. Higher array voltage means lower array current, so the savings in wiring costs can more than pay for the controller.

Charge controllers may also monitor battery temperature to prevent overheating. Some charge controller systems also display data, transmit data to remote displays, and data logging to track electric flow over time.

Integrated Charge Controller Circuitry

Circuitry that functions as a charge regulator controller may consist of several electrical components, or may be encapsulated in a single microchip, an integrated circuit (IC) usually called a charge controller IC or charge control IC.

Charge controller circuits are used for rechargeable electronic devices such as cell phones, laptop computers, portable audio players, and uninterruptible power supplies, as well as for larger battery systems found in electric vehicles and orbiting space satellites

Solar Pumping Inverters

Advanced solar pumping inverters convert DC voltage from the solar array into AC voltage to drive submersible pumps directly without the need for batteries or other energy storage devices. By utilizing MPPT (maximum power point tracking), solar pumping inverters regulate output frequency to control the speed of the pumps in order to save the pump motor from damage.

Solar pumping inverters usually have multiple ports to allow the input of DC current generated by PV arrays, one port to allow the output of AC voltage, and a further port for input from a water-level sensor.

Market

As of 2014, conversion efficiency for state-of-the-art solar converters reached more than 98 percent. While string inverters are used in residential to medium-sized commercial PV systems, central inverters cover the large commercial and utility-scale market. Market-share for central and string inverters are about 50 percent and 48 percent, respectively, leaving less than 2 percent to micro-inverters.

Inverter/converter market in 2014				
Type	**Power**	**Efficiency**[(a)]	**Market share**[(b)]	**Remarks**
String inverter	up to 100 kW_p[(c)]	98%	50%	Cost[(b)] €0.15 per watt-peak. Easy to replace.

Central inverter	above 100 kW_p	98.5%	48%	€0.10 per watt-peak. High reliability. Often sold along with a service contract.
Micro-inverter	module power range	90%–95%	1.5%	€0.40 per watt-peak. Ease of replacement concerns.
DC/DC converter Power optimizer	module power range	98.8%	N/A	€0.40 per watt-peak. Ease of replacement concerns. Inverter is still needed. About 0.75 GW_p installed in 2013.
Source: *data by IHS 2014, remarks by Fraunhofer ISE 2014, from: Photovoltaics Report, updated as per 8 September 2014, p. 35, PDF*				
Notes: [(a)]*best efficiencies displayed,* [(b)]*market-share and cost per watt are estimated,* [(c)]*kW_p = kilowatt-peak*				

Z-SOURCE INVERTER

A Z-source inverter is a type of power inverter, a circuit that converts direct current to alternating current. It functions as a buck-boost inverter without making use of DC-DC converter bridge due to its unique circuit topology.

Impedance (Z-) Source networks provide an efficient means of power conversion between source and load in a wide range of electric power conversion applications (dc–dc, dc–ac, ac–dc, ac– ac) , . Z-source-related research has grown rapidly since it was first proposed in 2002 by Prof. F. Z. Peng. The numbers of modifications and new Z-source topologies have grown exponentially. Improvements to the impedance networks by introducing coupled magnetics have also been lately proposed for achieving even higher voltage boosting, while using a shorter shoot-through time . They include the Γ-source, T-source, trans-Zsource, TZ-source, LCCT-Z-source (proposed in 2011 by **Dr Marek** Adamowicz and utilizing high frequency transformer connected in series with two dc-current-blocking capacitors) , high-frequency transformer-isolated, and Y-source networks. Among them, the Y-source network (proposed in year 2013 by Dr Yam P. Siwakoti) is more versatile and can in fact be viewed as the generic network, from which the Γ-source, T-source, and trans-Z-source networks are derived . The incommensurate properties of this network open a new horizon to researchers and engineers to explore, expand, and modify the circuit for a wide range of power conversion applications.

Types of Inverters

Inverters can be classified by their structure:

1. Single-phase inverter:

This type of inverter consists of two legs or two poles. (A pole is connection of two IGBTs where source of one and drain of other are connected and this common point is taken out).

2. Three-phase inverter:

This type of inverter consists of three legs or poles or four legs (three legs for phases and one for neutral).

But, inverters are also classified based on the type of input source. And they are,

1. Voltage-source inverter (VSI)

In this type of inverter, a constant voltage source acts as input to the inverter bridge. The constant voltage source is obtained by connecting a large capacitor across the DC source.

2. Current-source inverter (CSI)

In this type of inverter, a constant current source acts as input to the inverter bridge. The constant current source is obtained by connecting a large inductor in series the DC source.

Disadvantages

Typical inverters (VSI and CSI) have few disadvantages. They are listed as,

- Behave in a boost or buck operation only. Thus the obtainable output voltage range is limited, either smaller or greater than the input voltage.
- Vulnerable to EMI noise and the devices gets damaged in either open or short circuit conditions.
- The combined system of DC-DC boost converter and the inverter has lower reliability.
- The main switching device of VSI and CSI are not interchangeable.

Advantages of ZSI

The advantages of Z-source inverter are listed as follows,

- The source can be either a voltage source or a current source. The DC source of a ZSI can either be a battery, a diode rectifier or a thyristor converter, a fuel cell stack or a combination of these.
- The main circuit of a ZSI can either be the traditional VSI or the traditional CSI.
- Works as a buck-boost inverter.
- The load of a ZSC can either be inductive or capacitive or another Z-Source network.

REFERENCES

1. The Authoritative Dictionary of IEEE Standards Terms, Seventh Edition, IEEE Press, 2000,ISBN 0-7381-2601-2, page 588
2. http://www.solar-electric.com/lib/wind-sun/Pump-Inverter.pdf How to Choose an Inverter for an Independent Energy System
3. http://www.wpi.edu/Pubs/E-project/Available/E-project-042507-092653/

unrestricted/MQP_D_1_2.pdf

4. Barnes, Malcolm (2003). Practical variable speed drives and power electronics. Oxford: Newnes. p. 97. ISBN 0080473911.
5. Inverter selection
6. Tripp Lite: Power Inverter FAQ, http://www.tripplite.com/support/inverter-faq
7. "New and Cool: Variable Refrigerant Flow Systems". AIArchitect. American Institute of Architects. 2009-04-10. Retrieved 2013-08-06.
8. James, Hahn. "Modifi ed Sine-Wave Inverter Enhanced" (PDF). Power Electronics.
9. MIT open-courseware, Power Electronics, Spring 2007
10. Rodriguez, Jose; et al. (August 2002). "Multilevel Inverters: A Survey of Topologies, Controls, and Applications". IEEE Transactions on Industrial Electronics. IEEE. 49 (4): 724–738. doi:10.1109/TIE.2002.801052.
11. https://www.littleboxchallenge.com/
12. Owen, Edward L. (January–February 1996). "Origins of the Inverter". IEEE Industry Applications Magazine: History Department. IEEE. 2 (1): 64–66. doi:10.1109/2943.476602.
13. D. R. Grafham and J. C. Hey, editors, ed. (1972). SCR Manual (Fifth ed.). Syracuse, N.Y. USA: General Electric. pp. 236–239.
14. http://web.eecs.utk.edu/~tolbert/publications/ecce_2011_bailu.pdf
15. http://www05.abb.com/global/scot/scot271.nsf/veritydisplay/369669d5dd6e8e6ec1257ba500293166/$file/70-78%202m315_EN_72dpi.pdf
16. New electric vehicle technology packs more punch in smaller package, Phys.org, 14 October 2014
17. "Electricity storage: Location, location, location … and cost - Battery storage for transmission support in Alaska". eia.gov. Energy Information Administration (EIA). 2012. Retrieved July 23, 2012.
18. E-book on choosing a UPS topology based on application type "Avoiding Trap Doors Associated with Purchasing a UPS System" (PDF).
19. Solter, W. (2002), A new international UPS classification by IEC 62040-3, doi:10.1109/INTLEC.2002.1048709
20. Detailed explanation of UPS topologies "High-Availability Power Systems, Part I: UPS Internal Topology" (PDF). November 2000.
21. "Hydrogen Fuel Cell UPS".
22. "UPS On-Line Uninterruptible Power Supply Backup Power Source". Archived from the original on October 4, 2013.
23. "Hybrid Rotary UPS" (PDF). Archived from the original (PDF) on December 4, 2014.
24. http://h20000.www2.hp.com/bc/docs/support/SupportManual/c01173322/c01173322.pdf

25. My Ton (Ecos Consulting), Brian Fortenbery (EPRI), William Tschudi (LNBL) (January 2007). "DC Power for Improved Data Center Efficiency" (PDF). Lawrence Berkeley National Laboratory.
26. Active Power. "15 Seconds versus 15 Minutes: White Paper 107 Designing for High Availability" (PDF).
27. Tripp Lite: UPS Buying Guide, http://www.tripplite.com/products/ups-Buying-Guide
28. Detailed explanation of optimized N+1 configurations"Balancing Scalability and Reliability in the Critical Power System: When Does N + 1 Become Too Many + 1?" (PDF).
29. Detailed explanation of UPS redundancy options"High-Availability Power Systems, Part II: Redundancy Options" (PDF).
30. Refer to safety standard IEC 60950-22 or a local derivative according to location e.g. EN 60950-22 (Europe); UL 60950-22 (USA)
31. Raymond, Eric Steven. UPS HOWTO, section 3.3. The Linux Documentation Project, 2003–2007.
32. Generex. "Multi-XS User Manual" (PDF). Multi-XS is an active RS232 data switch, designed to handle serial communications of one UPS with up to 5 / 10 computers
33. APC AP9207 Share-UPS, User Manual, pp. 6–7, Port 1 is called the Advanced port because it supplies smart signaling, which provides the advanced capabilities available to a server running PowerChute plus software. The Advanced port provides full access to the Computer Interface port of the UPS. Ports 2–8 on the rear panel of Share-UPS are called Basic ports because they supply simple UPS signaling for On Battery and Low Battery conditions in the UPS. "Share-UPS User Manual" (PDF). Archived from the original (PDF) on April 24, 2012. Retrieved November 14, 2011.
34. An example of an Ethernet UPS controller: Liebert IntelliSlot Web Card Communications Interface Card
35. APC Application Note #67 "APC Network Management Card Security Implementation" (PDF). Archived from the original (PDF) on April 24, 2012. Retrieved November 14, 2011.
36. "How to calculate battery run-time". PowerStream Technologies. Retrieved 2010-04-26.
37. Saslow, Wayne M. (2002). Electricity, Magnetism, and Light. Toronto: Thomson Learning. pp. 302–4. ISBN 0-12-619455-6.
38. Peter M. Curtis (2011). Maintaining Mission Critical Systems in a 24/7 Environment. Wiley. pp. 261–262. ISBN 9781118041628.
39. Michael F. Hordeski (2005). Emergency and backup power sources: preparing for blackouts and brownouts. The Fairmont Press, Inc. ISBN 9780881734850.
40. Leonardo Energy. "Maintenance Manager's Guide, Section 2.1". Retrieved

August 1, 2012.

41. APC Inc. "Knowledgebase article: What is the expected life of my APC UPS battery?, Answer ID 8301".
42. "Maintaining and Testing Your UPS System to Ensure Continuous Power, Section: Maintaining a Battery Bank". The Data Center Journal.
43. BTECH Inc, BTECH's Focus – Predicting Battery Failure and Installation Manual, page 18, showing sensor wires for each cell/battery on a battery string, and also note that the current transducer sensors to detect cross-string series-parallel current recirculation.
44. mpoweruk.com, Battery and Energy Technologies, Cell Balancing, Woodbank Communications Ltd, Chester, UK.
45. datapowermonitoring.com, Battery Asset Management: VRLA ageing characteristics, Bart Cotton, founder and CEO, Data Power Monitoring Corporation, Batteries International, Jan 2005 Archived April 6, 2013, at the Wayback Machine.
46. Solar Cells and their Applications Second Edition, Lewis Fraas, Larry Partain, Wiley, 2010, ISBN 978-0-470-44633-1 , Section10.2.
47. "Invert your thinking: Squeezing more power out of your solar panels". scientificamerican.com. Retrieved 2011-06-09.
48. Comparison of Photovoltaic Array Maximum Power Point Tracking Techniques
49. Benanti, Travis L.; Venkataraman, D. (25 April 2005). "Organic Solar Cells: An Overview Focusing on Active Layer Morphology" (PDF). Photosynthesis Research. 87 (1): 77. doi:10.1007/s11120-005-6397-9. Retrieved 27 August 2013.
50. "Evaluation of Micro Controller Based Maximum Power Point Tracking Methods Using dSPACE Platform" (PDF). itee.uq.edu.au. Retrieved 2011-06-14.
51. "Comparative Study of Maximum Power Point Tracking Algorithms". doi:10.1002/pip.459.
52. http://dspace.mit.edu/bitstream/handle/1721.1/61308/702674669.pdf
53. "A SIDE-BY-SIDE COMPARISON OF MICRO AND CENTRAL INVERTERS IN SHADED AND UNSHADED CONDITIONS" (PDF). Retrieved 27 August 2013.
54. Photovoltaics: Design and Installation Manual. Newsociety Publishers. 2004. p. 80.
55. "Summary Report on the DOE High-tech Inverter Workshop" (PDF). Sponsored by the US Department of Energy, prepared by McNeil Technologies. eere.energy.gov. Retrieved 2011-06-10.
56. Fraunhofer ISE report , archived version as per September 2014 (archived PDF)

CHAPTER 6

AC to AC Convertors

Chapter Contents

An AC to AC converter simply means a device which can convert AC waveforms to another AC waveform. The output voltage and frequency can be managed according to requirements and usage. AC voltage transformers are thyristor based transformers which are used to convert fixed voltage to any required voltage without any change in frequency. Transformers that are used to converts AC power of one frequency to another different frequency are known as Cycloconverters.

INTRODUCTION

A solid-state AC-AC converter converts an AC waveform to another AC waveform, where the output voltage and frequency can be set arbitrarily.

CATEGORIES

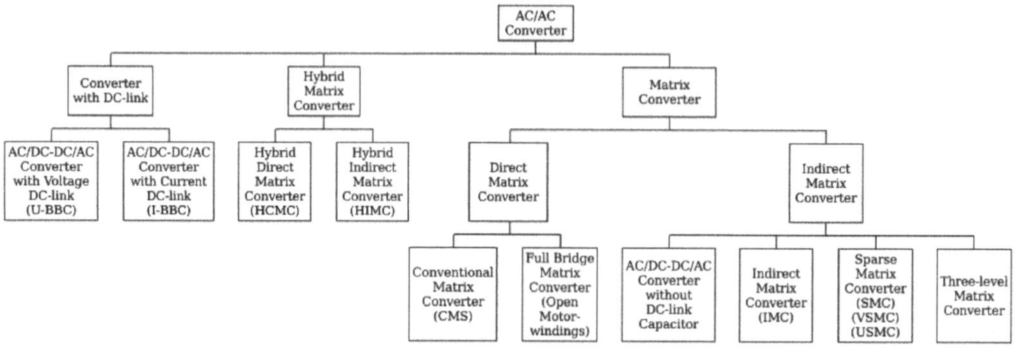

Classification of three-phase AC-AC converter circuits.

AC-AC converters can be categorized as follows:

- Indirect AC-AC (or AC/DC-AC) converters (i.e., with rectifier, DC link and inverter)
- Cycloconverters
- Hybrid matrix converters
- Matrix converters (MC).

DC LINK CONVERTERS

There are two types of converters with DC link:

- Voltage-source inverter (VSI) converters: In VSI converters, the rectifier consists of a diode-bridge and the DC link consists of a shunt capacitor.
- Current-source inverter (CSI) converters: In CSI converters, the rectifer consists of a phase-controlled switching device bridge and the DC link consists of 1 or 2 series inductors between one or both legs of the connection between rectifier and inverter.

Any dynamic braking operation required for the motor can be realized by means of braking DC chopper and resistor shunt connected across the rectifier. Alternatively, an anti-parallel thyristor bridge must be provided in the rectifier section to feed energy back into the AC line. Such phase-controlled thyristor-based rectifiers however have higher AC line distortion and lower power factor at low load than diode-based rectifiers.

An AC-AC converter with approximately sinusoidal input currents and bidirectional power flow can be realized by coupling a pulse-width modulation (PWM) rectifier and a PWM inverter to the DC-link. The DC-link quantity is then impressed by an energy storage element that is common to both stages, which is a capacitor C for the voltage DC-link or an inductor L for the current DC-link. The PWM rectifier is controlled in a way that a sinusoidal AC line current is drawn, which is in phase or anti-phase (for energy feedback) with the corresponding AC line phase voltage.

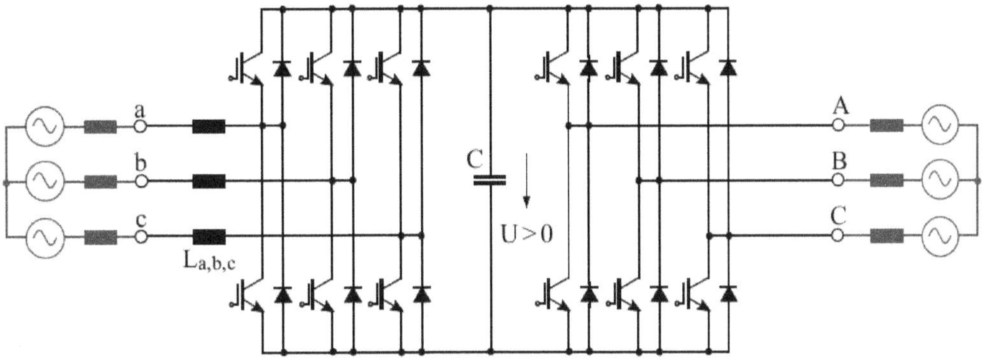

Topology of (regenerative) voltage-source inverter AC/DC-AC converter

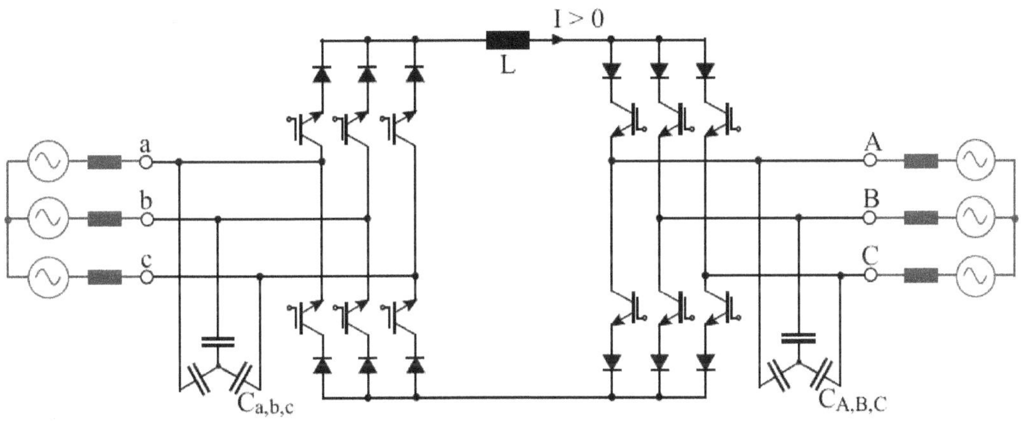

Topology of current-source inverter AC/DC-AC converter

Due to the DC-link storage element, there is the advantage that both converter stages are to a large extent decoupled for control purposes. Furthermore, a constant, AC line independent input quantity exists for the PWM inverter stage, which results in high utilization of the converter's power capability. On the other hand, the DC-link energy storage element has a relatively large physical volume, and when electrolytic capacitors are used, in the case of a voltage DC-link, there is potentially a reduced system lifetime.

MATRIX CONVERTERS

In order to achieve higher power density and reliability, it makes sense to consider Matrix Converters that achieve three-phase AC-AC conversion without any intermediate energy storage element. Conventional Direct Matrix Converters perform voltage and current conversion in one single stage.

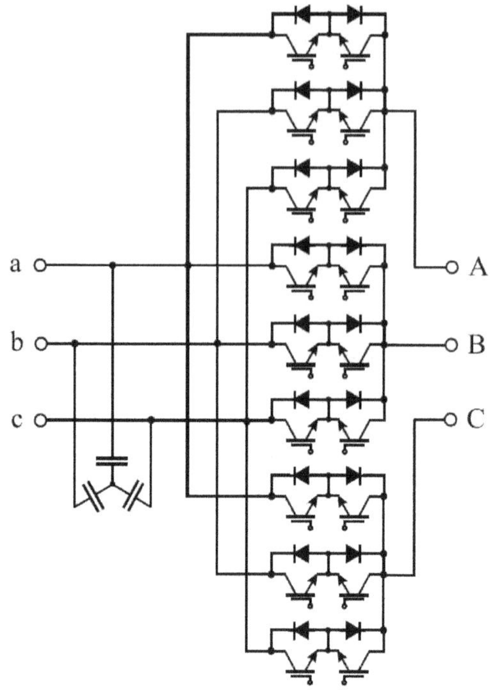

Topology of the Conventional Direct Matrix Converter

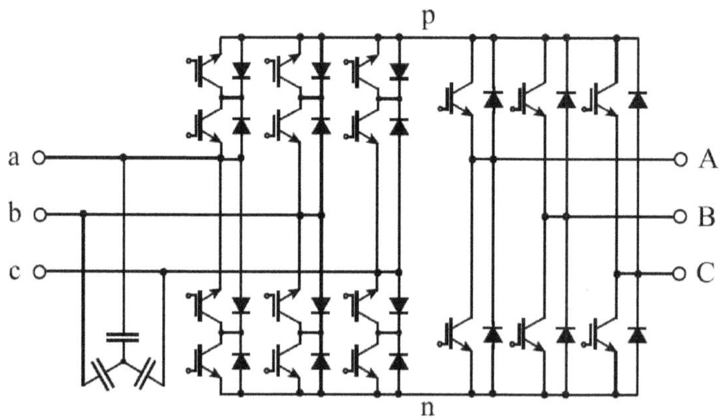

Topology of the indirect matrix converter

There is the alternative option of indirect energy conversion by employing the Indirect Matrix Converter or the Sparse matrix converter which was invented by Prof. Johann W. Kolar from the ETH Zurich. As with the DC-link based VSI and CSI controllers, separate stages are provided for voltage and current conversion, but the DC-link has no intermediate storage element. Generally, by employing matrix converters, the storage element in the DC-link is eliminated at the cost of a larger number of semiconductors.

Matrix converters are often seen as a future concept for variable speed drives technology, but despite intensive research over the decades they have until now only achieved low industrial penetration. However, citing recent availability of low-cost, high performance semiconductors, one larger drive manufacturer has over past few years been actively promoting matrix converters.

TRANSFORMER

A transformer is an electrical device that transfers electrical energy between two or more circuits through electromagnetic induction. Electromagnetic induction produces an electromotive force within a conductor which is exposed to time varying magnetic fields. Transformers are used to increase or decrease the alternating voltages in electric power applications.

A varying current in the transformer's primary winding creates a varying magnetic flux in the transformer core and a varying field impinging on the transformer's secondary winding. This varying magnetic field at the secondary winding induces a varying electromotive force (EMF) or voltage in the secondary winding due to electromagnetic induction. Making use of Faraday's Law (discovered in 1831) in conjunction with high magnetic permeability core properties, transformers can be designed to efficiently change AC voltages from one voltage level to another within power networks.

Since the invention of the first constant constant-potential transformer transformer in 1885, transformers have become essential for the transmission, distribution, and utilization of alternating current electrical energy. A wide range of transformer designs is encountered in electronic and electric power applications. Transformers range in size from RF transformers less than a cubic centimeter in volume to units interconnecting the power grid weighing hundreds of tons.

Pole-mounted distribution transformer with center-tapped secondary winding used to provide 'split-phase' power for residential and light commercial service, which in North America is typically rated 120/240 V.

Basic Principles

Ideal Transformer

For simplification or approximation purposes, it is very common to analyze the transformer as an ideal transformer model as presented in the two images. An ideal transformer is a theoretical, linear transformer that is lossless and perfectly coupled; that is, there are no energy losses and flux is completely confined within the magnetic core. Perfect coupling implies infinitely high core magnetic permeability and winding inductances and zero net magnetomotive force.

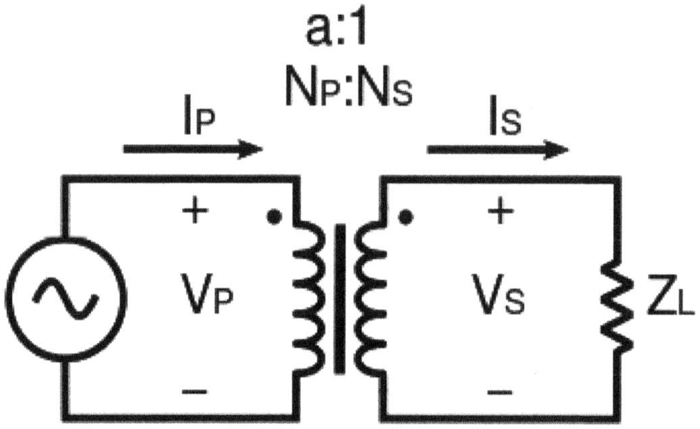

Ideal transformer connected with source V_P on primary and load impedance Z_L on secondary, where $0 < Z_L < \infty$.

A varying current in the transformer's primary winding creates a varying magnetic flux in the transformer core and a varying magnetic field impinging on the secondary winding. This varying magnetic field at the secondary winding induces a varying EMF or voltage in the secondary winding due to electromagnetic induction. The primary and secondary windings are wrapped around a core of infinitely high magnetic permeability so that all of the magnetic flux passes through both the primary and secondary windings. With a voltage source connected to the primary winding and load impedance connected to the secondary winding, the transformer currents flow in the indicated directions.

According to Faraday's Law, since the same magnetic flux passes through both the primary and secondary windings in an ideal transformer, a voltage is induced in each winding, according to eq. (1) in the secondary winding case, according to eq. (2) in the primary winding case. The primary EMF is sometimes termed counter EMF. This is in accordance with Lenz's law, which states that induction of EMF always opposes development of any such change in magnetic field.

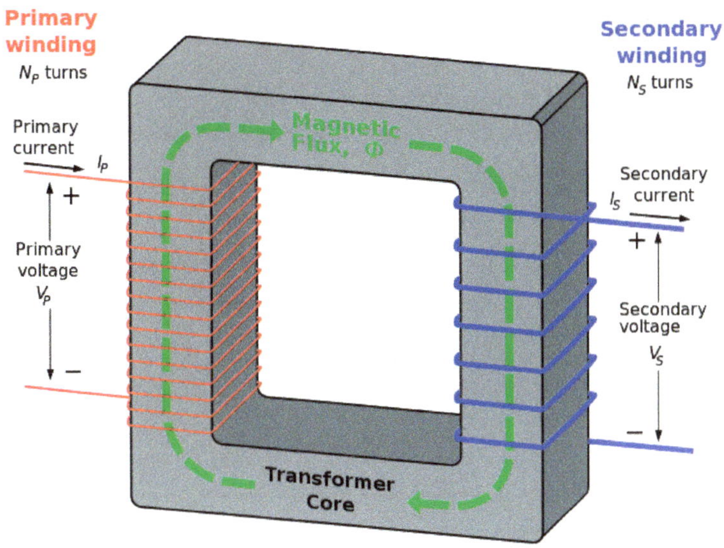

Ideal transformer and induction law

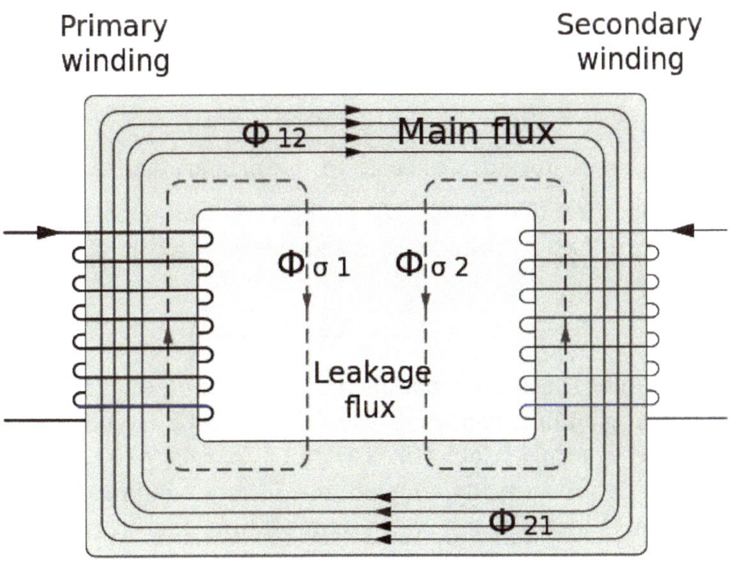

Leakage flux of a transformer.

The transformer winding voltage ratio is thus shown to be directly proportional to the winding turns ratio according to eq. (3).common usage having evolved over time from 'turn ratio' to 'turns ratio'. However, some sources use the inverse definition.

According to the law of conservation of energy, any load impedance connected to the ideal transformer's secondary winding results in conservation of apparent, real and reactive power consistent with eq. (4).

The ideal transformer identity shown in eq. (5) is a reasonable approximation for the typical commercial transformer, with voltage ratio and winding turns ratio both being inversely proportional to the corresponding current ratio.

By Ohm's law and the ideal transformer identity:

- the secondary circuit load impedance can be expressed as eq. (6)
- the apparent load impedance *referred* to the primary circuit is derived in eq. (7) to be equal to the turns ratio squared times the secondary circuit load impedance.

Real Transformer

Deviations from Ideal

The ideal transformer model neglects the following basic linear aspects in real transformers:

a) Core losses, collectively called magnetizing current losses, consisting of
   - Hysteresis losses due to nonlinear application of the voltage applied in the transformer core, and
   - Eddy current losses due to joule heating in the core that are proportional to the square of the transformer's applied voltage.

b) Whereas windings in the ideal model have no resistances and infinite inductances, the windings in a real transformer have finite non-zero resistances and inductances associated with:
   - Joule losses due to resistance in the primary and secondary windings
   - Leakage flux that escapes from the core and passes through one winding only resulting in primary and secondary reactive impedance.

Leakage Flux

The ideal transformer model assumes that all flux generated by the primary winding links all the turns of every winding, including itself. In practice, some flux traverses paths that take it outside the windings. Such flux is termed *leakage flux*, and results in leakage inductance in series with the mutually coupled transformer windings. Leakage flux results in energy being alternately stored in and discharged from the magnetic fields with each cycle of the power supply. It is not directly a power loss, but results in inferior voltage regulation, causing the secondary voltage not to be directly proportional to the primary voltage, particularly under heavy load. Transformers are therefore normally designed to have very low leakage inductance.

In some applications increased leakage is desired, and long magnetic paths, air gaps, or magnetic bypass shunts may deliberately be introduced in a transformer design to limit the short-circuit current it will supply.< Leaky transformers may be used to supply loads that exhibit negative resistance, such as electric arcs, mercury vapor lamps, and neon signs or for safely handling loads that become periodically short-circuited such as electric arc welders.

Air gaps are also used to keep a transformer from saturating, especially audio-frequency transformers in circuits that have a DC component flowing in the windings.

Knowledge of leakage inductance is also useful when transformers are operated in parallel. It can be shown that if the percent impedance and associated winding leakage reactance-to-resistance (*X*/*R*) ratio of two transformers were hypothetically exactly the same, the transformers would share power in proportion to their respective volt-ampere ratings (e.g. 500 kVA unit in parallel with 1,000 kVA unit, the larger unit would carry twice the current). However, the impedance tolerances of commercial transformers are significant. Also, the Z impedance and X/R ratio of different capacity transformers tends to vary, corresponding 1,000 kVA and 500 kVA units' values being, to illustrate, respectively, $Z \approx 5.75\%$, $X/R \approx 3.75$ and $Z \approx 5\%$, $X/R \approx 4.75$.

Equivalent Circuit

Referring to the diagram, a practical transformer's physical behavior may be represented by an equivalent circuit model, which can incorporate an ideal transformer.

Winding joule losses and leakage reactances are represented by the following series loop impedances of the model:

- Primary winding: R_P, X_P
- Secondary winding: R_S, X_S.

In normal course of circuit equivalence transformation, R_S and X_S are in practice usually referred to the primary side by multiplying these impedances by the turns ratio squared, $(N_P/N_S)^2 = a^2$.

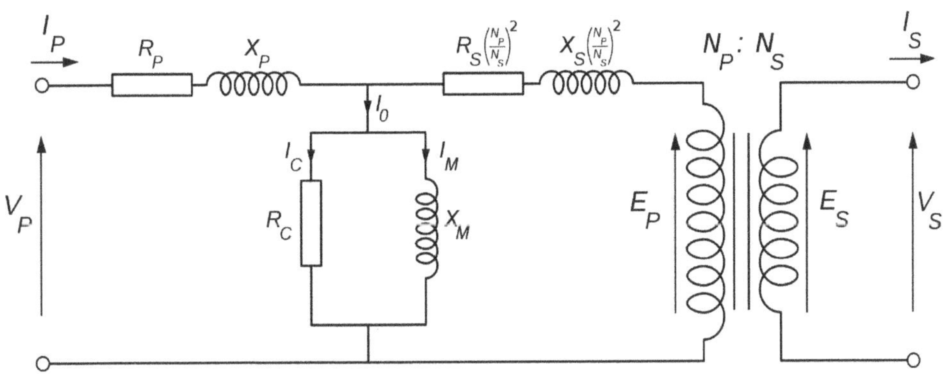

Real transformer equivalent circuit.

Core loss and reactance is represented by the following shunt leg impedances of the model:

- Core or iron losses: R_C

- Magnetizing reactance: X_M.

R_C and X_M are collectively termed the *magnetizing branch* of the model.

Core losses are caused mostly by hysteresis and eddy current effects in the core and are proportional to the square of the core flux for operation at a given frequency. The finite permeability core requires a magnetizing current I_M to maintain mutual flux in the core. Magnetizing current is in phase with the flux, the relationship between the two being non-linear due to saturation effects. However, all impedances of the equivalent circuit shown are by definition linear and such non-linearity effects are not typically reflected in transformer equivalent circuits. With sinusoidal supply, core flux lags the induced EMF by 90°. With open-circuited secondary winding, magnetizing branch current I_0 equals transformer no-load current.

Instrument transformer, with polarity dot and X1 markings on LV side terminal.

The resulting model, though sometimes termed 'exact' equivalent circuit based on linearity assumptions, retains a number of approximations. Analysis may be simplified by assuming that magnetizing branch impedance is relatively high and relocating the branch to the left of the primary impedances. This introduces error but allows combination of primary and referred secondary resistances and reactances by simple summation as two series impedances.

Transformer equivalent circuit impedance and transformer ratio parameters can be derived from the following tests: open-circuit test, short-circuit test, winding resistance test, and transformer ratio test.

Basic Transformer Parameters and Construction

Polarity

A dot convention is often used in transformer circuit diagrams, nameplates or terminal markings to define the relative polarity of transformer windings. Positively increasing instantaneous current entering the primary winding's dot end induces positive polarity voltage at the secondary winding's dot end.

Effect of Frequency

By Faraday's Law of induction shown in eq. (1) and (2), transformer EMFs vary according to the derivative of flux with respect to time. The ideal transformer's core behaves linearly with time for any non-zero frequency. Flux in a real transformer's core behaves non-linearly in relation to magnetization current as the instantaneous flux increases beyond a finite linear range resulting in magnetic saturation associated with increasingly large magnetizing current, which eventually leads to transformer overheating.

The EMF of a transformer at a given flux density increases with frequency. By operating at higher frequencies, transformers can be physically more compact because a given core is able to transfer more power without reaching saturation and fewer turns are needed to achieve the same impedance. However, properties such as core loss and conductor skin effect also increase with frequency. Aircraft and military equipment employ 400 Hz power supplies which reduce core and winding weight. Conversely, frequencies used for some railway electrification systems were much lower (e.g. 16.7 Hz and 25 Hz) than normal utility frequencies (50–60 Hz) for historical reasons concerned mainly with the limitations of early electric traction motors. Consequently, the transformers used to step-down the high overhead line voltages (e.g. 15 kV) were much larger and heavier for the same power rating than those required for the higher frequencies.

Operation of a transformer at its designed voltage but at a higher frequency than intended will lead to reduced magnetizing current. At a lower frequency, the magnetizing current will increase. Operation of a transformer at other than its design frequency may require assessment of voltages, losses, and cooling to establish if safe operation is practical. For example, transformers may need to be equipped with 'volts per hertz' over-excitation, ANSI function 24, relays to protect the transformer from overvoltage at higher than rated frequency.

One example is in traction transformer s used for electric multiple unit and high-speed train service operating across regions with different electrical standards. The converter equipment and traction transformers have to accommodate different input frequencies and voltage (ranging from as high as 50 Hz down to 16.7 Hz and rated up to 25 kV) while being suitable for multiple AC asynchronous motor and DC converters and motors with varying harmonics mitigation filtering requirements.

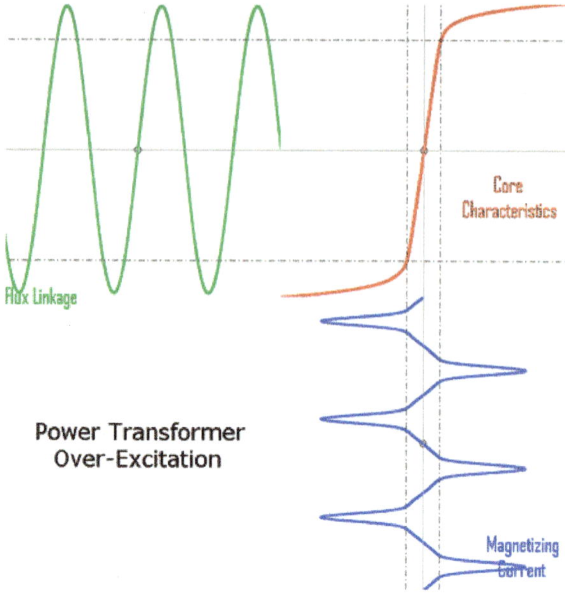

Power transformer over-excitation condition caused by decreased frequency; flux (green), iron core's magnetic characteristics (red) and magnetizing current (blue).

Large power transformers are vulnerable to insulation failure due to transient voltages with high-frequency components, such as caused in switching or by lightning.

At much higher frequencies the transformer core size required drops dramatically: a physically small and cheap transformer can handle power levels that would require a massive iron core at mains frequency. The development of switching power semiconductor devices and complex integrated circuits made switch-mode power supplies viable, to generate a high frequency from a much lower one (or DC), change the voltage level with a small transformer, and, if necessary, rectify the changed voltage.

Energy Losses

Real transformer energy losses are dominated by winding resistance joule and core losses. Transformers' efficiency tends to improve with increasing transformer capacity. The efficiency of typical distribution transformers is between about 98 and 99 percent.

As transformer losses vary with load, it is often useful to express these losses in terms of no-load loss, full-load loss, half-load loss, and so on. Hysteresis and eddy current losses are constant at all load levels and dominate overwhelmingly without load, while variable winding joule losses dominating increasingly as load increases. The no-load loss can be significant, so that even an idle transformer constitutes a drain on the electrical supply. Designing energy efficient transformers for lower loss requires a larger core, good-quality silicon steel, or even amorphous steel for the core and thicker wire, increasing initial cost. The choice of construction represents a trade-off between initial cost and operating cost.

Eddy Current Losses

Ferromagnetic materials are also good conductors and a core made from such a material also constitutes a single short-circuited turn throughout its entire length. Eddy currents therefore circulate within the core in a plane normal to the flux, and are responsible for resistive heating of the core material. The eddy current loss is a complex function of the square of supply frequency and inverse square of the material thickness. Eddy current losses can be reduced by making the core of a stack of plates electrically insulated from each other, rather than a solid block; all transformers operating at low frequencies use laminated or similar cores.

Magnetostriction Related Transformer Hum

Magnetic flux in a ferromagnetic material, such as the core, causes it to physically expand and contract slightly with each cycle of the magnetic field, an effect known as magnetostriction, the frictional energy of which produces an audible noise known as mains hum or transformer hum. This transformer hum is especially objectionable in transformers supplied at power frequencies and in high-frequency flyback transformers associated with television CRTs.

Stray Losses

Leakage inductance is by itself largely lossless, since energy supplied to its magnetic fields is returned to the supply with the next half-cycle. However, any leakage flux that intercepts nearby conductive materials such as the transformer's support structure will give rise to eddy currents and be converted to heat. There are also radiative losses due to the oscillating magnetic field but these are usually small.

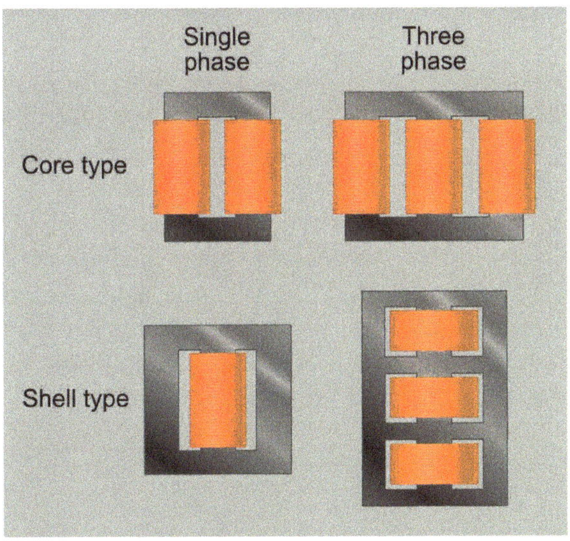

Core form = core type; shell form = shell type.

Mechanical Vibration and Audible Noise Transmission

In addition to magnetostriction, the alternating magnetic field causes fluctuating forces between the primary and secondary windings. This energy incites vibration transmission in interconnected metalwork, thus amplifying audible transformer hum.

Core Form and Shell Form Transformers

Closed-core transformers are constructed in 'core form' or 'shell form'. When windings surround the core, the transformer is core form; when windings are surrounded by the core, the transformer is shell form. Shell form design may be more prevalent than core form design for distribution transformer applications due to the relative ease in stacking the core around winding coils. Core form design tends to, as a general rule, be more economical, and therefore more prevalent, than shell form design for high voltage power transformer applications at the lower end of their voltage and power rating ranges (less than or equal to, nominally, 230 kV or 75 MVA). At higher voltage and power ratings, shell form transformers tend to be more prevalent. Shell form design tends to be preferred for extra-high voltage and higher MVA applications because, though more labor-intensive to manufacture, shell form transformers are characterized as having inherently better kVA-to-weight ratio, better short-circuit strength characteristics and higher immunity to transit damage.

Construction

Cores

Laminated Steel Cores

Transformers for use at power or audio frequencies typically have cores made of high permeability silicon steel. The steel has a permeability many times that of free space and the core thus serves to greatly reduce the magnetizing current and confine the flux to a path which closely couples the windings. Early transformer developers soon realized that cores constructed from solid iron resulted in prohibitive eddy current losses, and their designs mitigated this effect with cores consisting of bundles of insulated iron wires. Later designs constructed the core by stacking layers of thin steel laminations, a principle that has remained in use. Each lamination is insulated from its neighbors by a thin non-conducting layer of insulation. The universal transformer equation indicates a minimum cross-sectional area for the core to avoid saturation.

Laminated core transformer showing edge of laminations at top of photo.

The effect of laminations is to confine eddy currents to highly elliptical paths that enclose little flux, and so reduce their magnitude. Thinner laminations reduce losses, but are more laborious and expensive to construct. Thin laminations are generally used on high-frequency transformers, with some of very thin steel laminations able to operate up to 10 kHz.

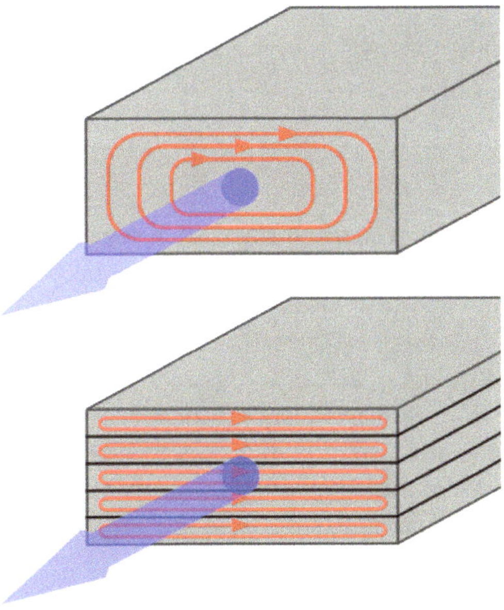

Laminating the core greatly reduces eddy-current losses.

One common design of laminated core is made from interleaved stacks of E-shaped steel sheets capped with I-shaped pieces, leading to its name of 'E-I transformer'. Such a design tends to exhibit more losses, but is very economical to manufacture. The cut-core or C-core type is made by winding a steel strip around a rectangular form and then bonding the layers together. It is then cut in two, forming two C shapes, and the core assembled by binding the two C halves together with a steel strap. They have the advantage that the flux is always oriented parallel to the metal grains, reducing reluctance.

A steel core's remanence means that it retains a static magnetic field when power is removed. When power is then reapplied, the residual field will cause a high inrush current until the effect of the remaining magnetism is reduced, usually after a few cycles of the applied AC waveform. Overcurrent protection devices such as fuses must be selected to allow this harmless inrush to pass. On transformers connected to long, overhead power transmission lines, induced currents due to geomagnetic disturbances during solar storms can cause saturation of the core and operation of transformer protection devices.

Distribution transformers can achieve low no-load losses by using cores made with low-loss high-permeability silicon steel or amorphous (non-crystalline) metal alloy. The higher initial cost of the core material is offset over the life of the transformer by its lower losses at light load.

Solid Cores

Powdered iron cores are used in circuits such as switch-mode power supplies that operate above mains frequencies and up to a few tens of kilohertz. These materials combine high magnetic permeability with high bulk electrical resistivity. For frequencies extending beyond the VHF band, cores made from non-conductive magnetic ceramic materials called ferrites are common. Some radio-frequency transformers also have movable cores (sometimes called 'slugs') which allow adjustment of the coupling coefficient (and bandwidth) of tuned radio-frequency circuits.

Toroidal Cores

Toroidal transformers are built around a ring-shaped core, which, depending on operating frequency, is made from a long strip of silicon steel or permalloy wound into a coil, powdered iron, or ferrite. A strip construction ensures that the grain boundaries are optimally aligned, improving the transformer's efficiency by reducing the core's reluctance. The closed ring shape eliminates air gaps inherent in the construction of an E-I core. The cross-section of the ring is usually square or rectangular, but more expensive cores with circular cross-sections are also available. The primary and secondary coils are often wound concentrically to cover the entire surface of the core. This minimizes the length of wire needed and provides screening to minimize the core's magnetic field from generating electromagnetic interference.

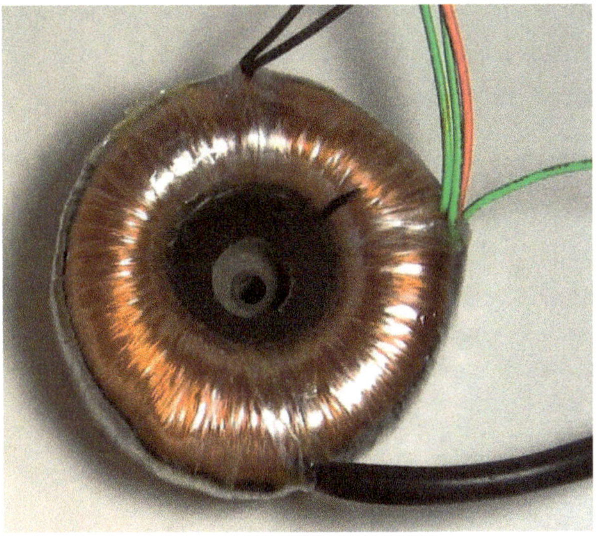

Small toroidal core transformer.

Toroidal transformers are more efficient than the cheaper laminated E-I types for a similar power level. Other advantages compared to E-I types, include smaller size (about half), lower weight (about half), less mechanical hum (making them superior in audio amplifiers), lower exterior magnetic field (about one tenth), low off-load losses (making them more efficient in standby circuits), single-bolt mounting, and greater choice of shapes. The main disadvantages are higher cost and limited power capacity. Because of the lack of a residual gap in the magnetic path, toroidal transformers also tend to exhibit higher inrush current, compared to laminated E-I types.

Ferrite toroidal cores are used at higher frequencies, typically between a few tens of kilohertz to hundreds of megahertz, to reduce losses, physical size, and weight of inductive components. A drawback of toroidal transformer construction is the higher labor cost of winding. This is because it is necessary to pass the entire length of a coil winding through the core aperture each time a single turn is added to the coil. As a consequence, toroidal transformers rated more than a few kVA are uncommon. Relatively few toroids are offered with power ratings above 10 kVA, and practically none above 25 kVA. Small distribution transformers may achieve some of the benefits of a toroidal core by splitting it and forcing it open, then inserting a bobbin containing primary and secondary windings.

Air Cores

A physical core is not an absolute requisite and a functioning transformer can be produced simply by placing the windings near each other, an arrangement termed an 'air-core' transformer. The air which comprises the magnetic circuit is essentially lossless, and so an air-core transformer eliminates loss due to hysteresis in the core material. The leakage inductance is inevitably high, resulting in very poor regulation, and so such designs are unsuitable for use in power distribution. They have however very high bandwidth, and are frequently employed in radio-frequency applications, for which a satisfactory coupling coefficient is maintained by carefully overlapping the primary and secondary windings. They're also used for resonant transformers such as Tesla coils where they can achieve reasonably low loss in spite of the high leakage inductance.

Windings

High-frequency transformers operating in the tens to hundreds of kilohertz often have windings made of braided Litz wire to minimize the skin-effect and proximity effect losses. Large power transformers use multiple-stranded conductors as well, since even at low power frequencies non-uniform distribution of current would otherwise exist in high-current windings. Each strand is individually insulated, and the strands are arranged so that at certain points in the winding, or throughout the whole winding, each portion occupies different relative positions in the complete conductor. The transposition equalizes the current flowing in each strand of the conductor, and reduces eddy current losses in the winding itself. The stranded conductor is also more flexible than a solid conductor of similar size, aiding manufacture.

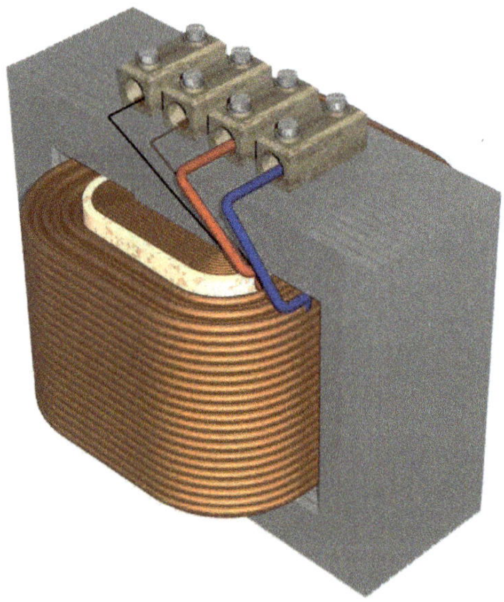

Windings are usually arranged concentrically to minimize flux leakage.

The windings of signal transformers minimize leakage inductance and stray capacitance to improve high-frequency response. Coils are split into sections, and those sections interleaved between the sections of the other winding.

Power-frequency transformers may have taps at intermediate points on the winding, usually on the higher voltage winding side, for voltage adjustment. Taps may be manually reconnected, or a manual or automatic switch may be provided for changing taps. Automatic on-load tap changers are used in electric power transmission or distribution, on equipment such as arc furnace transformers, or for automatic voltage regulators for sensitive loads. Audio-frequency transformers, used for the distribution of audio to public address loudspeakers, have taps to allow adjustment of impedance to each speaker. A center-tapped transformer is often used in the output stage of an audio power amplifier in a push-pull circuit. Modulation transformers in AM transmitters are very similar.

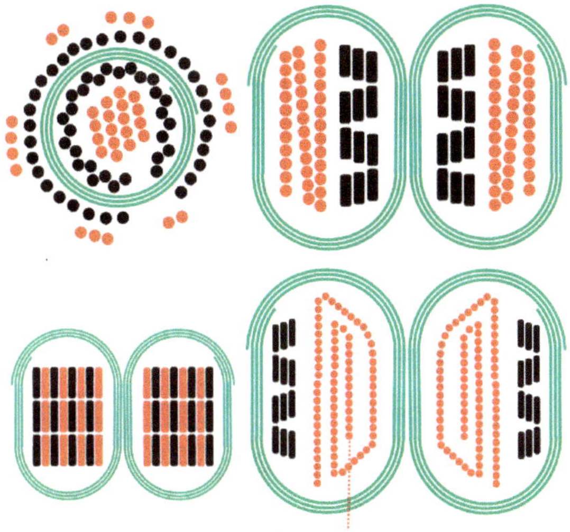

Cut view through transformer windings.

Dry-type transformer winding insulation systems can be either of standard open-wound 'dip-and-bake' construction or of higher quality designs that include vacuum pressure impregnation (VPI), vacuum pressure encapsulation (VPE), and cast coil encapsulation processes. In the VPI process, a combination of heat, vacuum and pressure is used to thoroughly seal, bind, and eliminate entrained air voids in the winding polyester resin insulation coat layer, thus increasing resistance to corona. VPE windings are similar to VPI windings but provide more protection against environmental effects, such as from water, dirt or corrosive ambients, by multiple dips including typically in terms of final epoxy coat.

Regarding image at top captioned, Cut view of transformer windings:

The conducting material used for the windings depends upon the application, but in all cases the individual turns must be electrically insulated from each other to ensure that the current travels throughout every turn. For small power and signal transformers, in which currents are low and the potential difference between adjacent turns is small, the coils are often wound from enamelled magnet wire, such as Formvar wire. Larger power transformers operating at high voltages may be wound with copper rectangular strip conductors insulated by oil-impregnated paper and blocks of pressboard.

Legend

White: insulator.

Green spiral: Grain oriented silicon steel.

Black: Primary winding made of oxygen-free copper.

Red: Secondary winding.

Cooling

To place the cooling problem in perspective, the accepted rule of thumb is that the life expectancy of insulation in all electrics, including all transformers, is halved for about every 7 °C to 10 °C increase in operating temperature, this life expectancy halving rule holding more narrowly when the increase is between about 7 °C to 8 °C in the case of transformer winding cellulose insulation.

Cutaway view of liquid-immersed construction transformer. The conservator (reservoir) at top provides liquid-to-atmosphere isolation as coolant level and temperature changes. The walls and fins provide required heat dissipation balance.

Small dry-type and liquid-immersed transformers are often self-cooled by natural convection and radiation heat dissipation. As power ratings increase, transformers are often cooled by forced-air cooling, forced-oil cooling, water-cooling, or combinations of these. Large transformers are filled with transformer oil that both cools and insulates the windings. Transformer oil is a highly refined mineral oil that cools the windings and insulation by circulating within the transformer tank. The mineral oil and paper insulation system has been extensively studied and used for more than 100 years. It is estimated that 50% of power transformers will survive 50 years of use, that the average age of failure of power transformers is about 10 to 15 years, and that about 30% of power transformer failures are due to insulation and overloading failures. Prolonged operation at elevated temperature degrades insulating properties of winding insulation and dielectric coolant, which not only shortens transformer life but can ultimately lead to catastrophic transformer failure. With a great body of empirical study as a guide, transformer oil testing including dissolved gas analysis provides valuable maintenance information. This underlines the need to monitor, model, forecast and manage oil and winding conductor insulation temperature conditions under varying, possibly difficult, power loading conditions.

Building regulations in many jurisdictions require indoor liquid-filled transformers to either use dielectric fluids that are less flammable than oil, or be installed in fire-resistant rooms. Air-cooled dry transformers can be more economical where they eliminate the cost of a fire-resistant transformer room.

The tank of liquid filled transformers often has radiators through which the liquid coolant circulates by natural convection or fins. Some large transformers employ electric fans for forced-air cooling, pumps for forced-liquid cooling, or have heat exchangers for water-cooling. An oil-immersed transformer may be equipped with a Buchholz relay, which, depending on severity of gas accumulation due to internal arcing, is used to either alarm or de-energize the transformer. Oil-immersed transformer installations usually include fire protection measures such as walls, oil containment, and fire-suppression sprinkler systems.

Polychlorinated biphenyls have properties that once favored their use as a dielectric coolant, though concerns over their environmental persistence led to a widespread ban on their use. Today, non-toxic, stable silicone-based oils, or fluorinated hydrocarbons may be used where the expense of a fire-resistant liquid offsets additional building cost for a transformer vault. PCBs for new equipment were banned in 1981 and in 2000 for use in existing equipment in United Kingdom Legislation enacted in Canada between 1977 and 1985 essentially bans PCB use in transformers manufactured in or imported into the country after 1980, the maximum allowable level of PCB contamination in existing mineral oil transformers being 50 ppm.

Some transformers, instead of being liquid-filled, have their windings enclosed in sealed, pressurized tanks and cooled by nitrogen or sulfur hexafluoride gas.

Experimental power transformers in the 500-to-1,000 kVA range have been built with liquid nitrogen or helium cooled superconducting windings, which eliminates winding losses without affecting core losses.

Insulation Drying

Construction of oil-filled transformers requires that the insulation covering the windings be thoroughly dried of residual moisture before the oil is introduced. Drying is carried out at the factory, and may also be required as a field service. Drying may be done by circulating hot air around the core, or by vapor-phase drying (VPD) where an evaporated solvent transfers heat by condensation on the coil and core.

For small transformers, resistance heating by injection of current into the windings is used. The heating can be controlled very well, and it is energy efficient. The method is called low-frequency heating (LFH) since the current used is at a much lower frequency than that of the power grid, which is normally 50 or 60 Hz. A lower frequency reduces the effect of inductance, so the voltage required can be reduced. The LFH drying method is also used for service of older transformers.

Bushings

Larger transformers are provided with high-voltage insulated bushings made of polymers or porcelain. A large bushing can be a complex structure since it must provide careful control of the electric field gradient without letting the transformer leak oil.

Classification Parameters

Transformers can be classified in many ways, such as the following:

- *Power capacity*: From a fraction of a volt-ampere (VA) to over a thousand MVA.
- *Duty of a transformer*: Continuous, short-time, intermittent, periodic, varying.
- *Frequency range*: Power-frequency, audio-frequency, or radio-frequency.
- *Voltage class*: From a few volts to hundreds of kilovolts.
- *Cooling type*: Dry and liquid-immersed – self-cooled, forced air-cooled; liquid-immersed – forced oil-cooled, water-cooled.
- *Circuit application*: Such as power supply, impedance matching, output voltage and current stabilizer or circuit isolation.
- *Utilization*: Pulse, power, distribution, rectifier, arc furnace, amplifier output, etc..
- *Basic magnetic form*: Core form, shell form, concentric, sandwich.
- *Constant-potential transformer descriptor*: Step-up, step-down, isolation.
- *General winding configuration*: By EIC vector group – various possible two-winding combinations of the phase designations delta, wye or star, and zigzag or interconnected star; other – autotransformer, Scott-T, zigzag grounding transformer winding.
- *Rectifier phase-shift winding configuration*: 2-winding, 6-pulse; 3-winding, 12-pulse; . . . n-winding, [n-1]*6-pulse; polygon; etc..

Types

Various specific electrical application designs require a variety of transformer types. Although they all share the basic characteristic transformer principles, they are customize in construction or electrical properties for certain installation requirements or circuit conditions.

- *Autotransformer*: Transformer in which part of the winding is common to both primary and secondary circuits, leading to increased efficiency, smaller size, and a higher degree of voltage regulation.
- *Capacitor voltage transformer*: Transformer in which capacitor divider is used to reduce high voltage before application to the primary winding.

- *Distribution transformer, power transformer*: International standards make a distinction in terms of distribution transformers being used to distribute energy from transmission lines and networks for local consumption and power transformers being used to transfer electric energy between the generator and distribution primary circuits.
- *Phase angle regulating transformer*: A specialized transformer used to control the flow of real power on three-phase electricity transmission networks.
- *Scott-T transformer*: Transformer used for phase transformation from three-phase to two-phase and vice versa.
- *Polyphase transformer*: Any transformer with more than one phase.
- *Grounding transformer*: Transformer used for grounding three-phase circuits to create a neutral in a three wire system, using a wye-delta transformer, or more commonly, a zigzag grounding winding.
- *Leakage transformer*: Transformer that has loosely coupled windings.
- *Resonant transformer*: Transformer that uses resonance to generate a high secondary voltage.
- *Audio transformer*: Transformer used in audio equipment.
- *Output transformer*: Transformer used to match the output of a valve amplifier to its load.
- *Instrument transformer*: Potential or current transformer used to accurately and safely represent voltage, current or phase position of high voltage or high power circuits.
- *Pulse transformer*: Specialized small-signal transformer used to transmit digital signaling while providing electrical isolation, commonly used in Ethernet computer networks as 10BASE-T, 100BASE-T and 1000BASE-T.

An electrical substation in Melbourne, Australia showing three of five 220 kV – 66 kV transformers, each with a capacity of 150 MVA

Applications

Transformers are used to increase (or step-up) voltage before transmitting electrical energy over long distances through wires. Wires have resistance which loses energy through joule heating at a rate corresponding to square of the current. By transforming power to a higher voltage transformers enable economical transmission of power and distribution. Consequently, transformers have shaped the electricity supply industry, permitting generation to be located remotely from points of demand. All but a tiny fraction of the world's electrical power has passed through a series of transformers by the time it reaches the consumer.

Transformer at the Limestone Generating Station in Manitoba, Canada.

Since the high voltages carried in the wires are significantly greater than what is needed in-home, transformers are also used extensively in electronic products to decrease (or step-down) the supply voltage to a level suitable for the low voltage circuits they contain. The transformer also electrically isolates the end user from contact with the supply voltage.

Signal and audio transformers are used to couple stages of amplifiers and to match devices such as microphones and record players to the input of amplifiers. Audio transformers allowed telephone circuits to carry on a two-way conversation over a single pair of wires. A balun transformer converts a signal that is referenced to ground to a signal that has balanced voltages to ground, such as between external cables and internal circuits. Transformers made to medical grade standards isolate the users from the direct current. These are found commonly used in conjunction with hospital beds, dentist chairs, and other medical lab equipment.

AUTOTRANSFORMER

An autotransformer (sometimes called *autostep down transformer*) is an electrical transformer with only one winding. The "auto" (Greek for "self") prefix refers to the single coil acting alone and not to any kind of automatic mechanism. In an autotransformer, portions of the same winding act as both the primary and secondary sides of the transformer. In contrast, an ordinary transformer has separate primary and secondary windings which are not electrically connected.

The winding has at least three taps where electrical connections are made. Since part of the winding does "double duty", autotransformers have the advantages of often being smaller, lighter, and cheaper than typical dual-winding transformers, but the disadvantage of not providing electrical isolation between primary and secondary circuits. Other advantages of autotransformers include lower leakage reactance, lower losses, lower excitation current, and increased VA rating for a given size and mass.

Autotransformers are often used to step up or step down voltages in the 110-115-120 V range and voltages in the 220-230-240 V range - for example, providing 110 V or 120 V (with taps) from 230 V input, allowing equipment designed for 100 or 120 V to be used with a 230 V supply. This allows US electrical equipment to be fed from the higher European voltage. Autotransformers can also be used to supply European 230 V appliances from a 100 or 120 V supply in countries outside Europe. In all cases the supply and the autotransformer must be correctly rated to supply the required power.

Operation

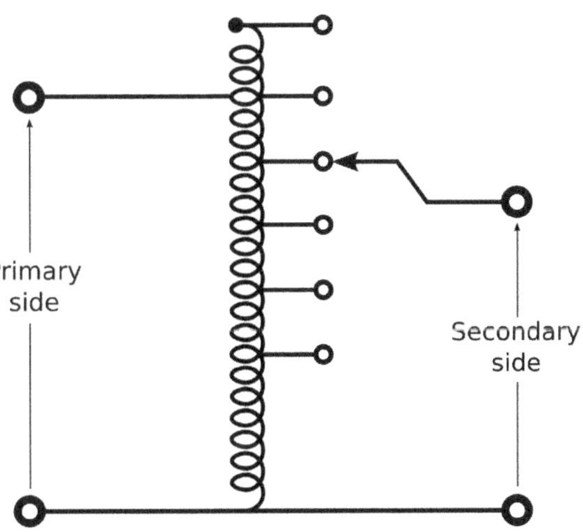

Single-phase tapped autotransformer with output voltage range of 40%–115% of input

An autotransformer has a single winding with two end terminals, and one or more terminals at intermediate tap points, or it is a transformer in which the primary and secondary coils have part of, or all of their turns in common. The primary voltage is applied across two of the terminals, and the secondary voltage taken from two terminals, almost always having one terminal in common with the primary voltage. The primary and secondary circuits therefore have a number of windings turns in common. Since the volts-per-turn is the same in both windings, each develops a voltage in proportion to its number of turns. In an autotransformer part of the current flows directly from the input to the output, and only part is transferred inductively, allowing a smaller, lighter, cheaper core to be used as well as requiring only a single winding. However the voltage and current ratio of autotransformers can be formulated the same as other two-winding transformers:

$$\frac{V_1}{V_2} = \frac{N_1}{N_2} = a$$

$(0<V_2<V_1)$

The ampere-turns provided by the upper half:

$$F_U = (N_1 - N_2)I_1 = (1 - \frac{1}{a})N_1 I_1$$

The ampere-turns provided by the lower half:

$$F_L = N_2(I_2 - I_1) = \frac{N_1}{a}(I_2 - I_1)$$

For ampere-turn balance, $F_U=F_L$:

$$(1 - \frac{1}{a})N_1 I_1 = \frac{N_1}{a}(I_2 - I_1)$$

Therefore:

$$\frac{I_1}{I_2} = \frac{1}{a}$$

One end of the winding is usually connected in common to both the voltage source and the electrical load. The other end of the source and load are connected to taps along the winding. Different taps on the winding correspond to different voltages, measured from the common end. In a step-down transformer the source is usually connected across the entire winding while the load is connected by a tap across only a portion of the winding. In a step-up transformer, conversely, the load is attached across the full winding while the source is connected to a tap across a portion of the winding.

As in a two-winding transformer, the ratio of secondary to primary voltages is equal to the ratio of the number of turns of the winding they connect to. For example, connecting the load between the middle and bottom of the autotransformer will reduce the voltage by 50%. Depending on the application, that portion of the winding used solely in the higher-voltage (lower current) portion may be wound with wire of a smaller gauge, though the entire winding is directly connected.

If one of the center-taps is used for the ground, then the autotransformer can be used as a balun to convert a balanced line (connected to the two end taps) to an unbalanced line (the side with the ground).

Limitations

An autotransformer does not provide electrical isolation between its windings as an ordinary transformer does; if the neutral side of the input is not at ground voltage, the neutral side of the output will not be either. A failure of the isolation of the windings of an autotransformer can result in full input voltage applied to the output. Also, a break in the part of the winding that is used as both primary and secondary will result in the transformer acting as an inductor in series with the load (which under light load conditions may result in near full input voltage being applied to the output). These are important safety considerations when deciding to use an autotransformer in a given application.

Because it requires both fewer windings and a smaller core, an autotransformer for power applications is typically lighter and less costly than a two-winding transformer, up to a voltage ratio of about 3:1; beyond that range, a two-winding transformer is usually more economical.

In three phase power transmission applications, autotransformers have the limitations of not suppressing harmonic currents and as acting as another source of ground fault currents. A large three-phase autotransformer may have a "buried" delta winding, not connected to the outside of the tank, to absorb some harmonic currents.

In practice, losses mean that both standard transformers and autotransformers are not perfectly reversible; one designed for stepping down a voltage will deliver slightly less voltage than required if it is used to step up. The difference is usually slight enough to allow reversal where the actual voltage level is not critical.

Like multiple-winding transformers, autotransformers use time-varying magnetic fields to transfer power. They require alternating currents to operate properly and will not function on direct current.

Applications

Power Transmission and Distribution

Autotransformers are frequently used in power applications to interconnect systems operating at different voltage classes, for example 132 kV to 66 kV for transmission. Another application in industry is to adapt machinery built (for example) for 480 V supplies to operate on a 600 V supply. They are also often used for providing conversions between the two common domestic mains voltage bands in the world (100 V—130 V and 200 V—250 V). The links between the UK 400 kV and 275 kV 'Super Grid' networks are normally three phase autotransformers with taps at the common neutral end.

On long rural power distribution lines, special autotransformers with automatic tap-changing equipment are inserted as voltage regulators, so that customers at the far end of the line receive the same average voltage as those closer to the source. The variable ratio of the autotransformer compensates for the voltage drop along the line.

A special form of autotransformer called a *zig zag* is used to provide grounding on three-phase systems that otherwise have no connection to ground. A zig-zag transformer provides a path for current that is common to all three phases (so-called *zero sequence* current).

Audio System

In audio applications, tapped autotransformers are used to adapt speakers to constant-voltage audio distribution systems, and for impedance matching such as between a low-impedance microphone and a high-impedance amplifier input.

Railways

In UK railway applications, it is common to power the trains at 25 kV AC. To increase the distance between electricity supply Grid feeder points they can be arranged to supply a 25-0-25 kV supply with the third wire (opposite phase) out of reach of the train's overhead collector pantograph. The 0 V point of the supply is connected to the rail while one 25 kV point is connected to the overhead contact wire. At frequent (about 10 km) intervals, an autotransformer links the contact wire to rail and to the second (antiphase) supply conductor. This system increases usable transmission distance, reduces induced interference into external equipment and reduces cost. A variant is occasionally seen where the supply conductor is at a different voltage to the contact wire with the autotransformer ratio modified to suit.

Variable Autotransformers

By exposing part of the winding coils and making the secondary connection through a sliding brush, a continuously variable turns ratio can be obtained, allowing for very smooth control of output voltage. The output voltage is not limited to the discrete voltages represented by actual number of turns. The voltage can be smoothly varied between turns as the brush has a relatively high resistance (compared with a metal contact) and the actual output voltage is a function of the relative area of brush in contact with adjacent windings. The relatively high resistance of the brush also prevents it from acting as a short circuited turn when it contacts two adjacent turns. Typically the primary connection connects to only a part of the winding allowing the output voltage to be varied smoothly from zero to above the input voltage and thus allowing the device to be used for testing electrical equipment at the limits of its specified voltage range.

A variable autotransformer, with a sliding-brush secondary connection and a toroidal core. Cover has been removed to show copper windings and brush.

Variable Transformer - part of Tektronix 576 Curve Tracer.

The output voltage adjustment can be manual or automatic. The manual type is applicable only for relatively low voltage and is known as a variable AC transformer (often referred to by the trademark name Variac). These are often used in repair shops for testing devices under different voltages or to simulate abnormal line voltages.

The type with automatic voltage adjustment can be used as automatic voltage regulator, to maintain a steady voltage at the customers' service during a wide range of line and load conditions. Another application is a lighting dimmer that doesn't produce the EMI typical of most thyristor dimmers.

Variac Trademark

From 1934 to 2002, Variac was a U.S. trademark of General Radio for a variable autotransformer intended to conveniently vary the output voltage for a steady AC input voltage. In 2004, Instrument Service Equipment applied for and obtained the *Variac* trademark for the same type of product. This word has become generic for hand-variable autotransformers in general.

VOLTAGE CONVERTER

A voltage converter is an electric power converter which changes the voltage of an electrical power source. It may be combined with other components to create a power supply.

AC and DC

AC voltage conversion uses a transformer. Conversion from one DC voltage to another requires electronic circuitry (electromechanical equipment was required before the development of semiconductor electronics). Mains power (called household current in the US) is universally AC.

Practical Voltage Converters

Mains Converters

A common use of the voltage converter is for a device that allows appliances made for the mains voltage of one geographical region to operate in an area with different voltage. Such a device may be called a *voltage converter*, *power converter*, *travel adapter*, etc. Most single phase alternating-current electrical outlets in the world supply power at 210–240 V or at 100–120 V. A transformer or autotransformer can be used; (auto) transformers are inherently reversible, so the same transformer can be used to step the voltage up, or step it down by the same ratio. Lighter and smaller devices can be made using electronic circuitry; reducing the voltage electronically is simpler and cheaper than increasing it. Small, inexpensive, *travel adapters* suitable for low-power devices such as electric shavers, but not, say, hairdriers, are available; travel adapters usually include plug-end adapters for the different standards used in different countries. A transformer would be used for higher power.

Transformers do not change the frequency of electricity; in many regions with 100–120 V, electricity is supplied at 60 Hz, and 210–240 V regions tend to use 50 Hz. This may affect operation of devices which depend on mains frequency (some audio turntables and mains-only electric clocks, etc., although modern equipment is less likely to depend upon mains frequency). Equipment with high-powered motors or internal transformers designed to operate at 60 Hz may overheat at 50 Hz even if the voltage supplied is correct.

Most mains-powered electrical equipment, though it may specify a single nominal voltage, actually has a range of tolerance above and below that point. Thus, devices usually can be used on either any voltage from approx. 100 to 120 V, or any voltage from approx. 210 to 240 V. In such cases, voltage converters need only be specified to convert any voltage within one range, to a voltage within the other, rather than separate converters being needed for all possible pairs of nominal voltages (110–220, 117–220, 110–230, etc.)

Converters for Devices

Mains Converters

Another requirement is to provide low-voltage electricity to a device from mains electricity; this would be done by what is usually called a *power supply*. Most modern electronic devices require between 1.5 and 24 volts DC; lower-powered devices at these voltages can often work either from batteries or mains. Some devices incorporate a power supply and are simply plugged into the mains. Others use an external power supply comprising either a transformer and rectifier, or electronic circuitry. Switched-mode power supplies have become widespread in the early twenty-first century; they are smaller and lighter than the once-universal transformer converters, and are often designed to work from AC mains at any voltage between 100 and 250 V. Additionally, because they are typically rectified to operate at a DC voltage, they are minimally affected by the frequency of the mains (50 vs 60 Hz). Details on operation are given in the article on power supplies.

Mobile Converters

Voltage converters can be used in vehicles with 12 V DC outlets. A simple voltage dropper can be used to reduce the voltage for low-power devices; if more than 12V is required, or for high-powered devices, a switched-mode power supply is used. The output will usually be DC in the range 1.5–24 V. Power supplies that output either 100–120 V AC or 210–240 V AC are available; they are called inverters, due to the conversion from DC to AC rather than the voltage change. The output frequency and waveform of an inverter may not accurately replicate that supplied by mains electricity, although this is not usually a problem.

Choosing a Converter

A converter to drive equipment should be specified to supply at least the actual watts or amperes used by the equipment; this is usually stated on a label on the equipment. A converter of higher rating is safe to use. Some equipment uses more power or current at startup; a 20% margin is usually adequate, although it may be included in the rated figure.

VOLTAGE REGULATOR

A voltage regulator is designed to automatically maintain a constant voltage level. A voltage regulator may be a simple "feed-forward" design or may include negative feedback control loops. It may use an electromechanical mechanism, or electronic components. Depending on the design, it may be used to regulate one or more AC or DC voltages.

Electronic voltage regulators are found in devices such as computer power supplies where they stabilize the DC voltages used by the processor and other elements. In automobile alternators and central power station generator plants, voltage regulators control the output of the plant. In an electric power distribution system, voltage regulators may be installed at a substation or along distribution lines so that all customers receive steady voltage independent of how much power is drawn from the line.

An integrated circuit voltage regulator in a TO-220 style package. Such devices are popular because they require few or no external components and provide the functions of pass element, voltage reference, and protection from overcurrent in one package.

Electronic Voltage Regulators

A simple voltage/current regulator can be made from a resistor in series with a diode (or series of diodes). Due to the logarithmic shape of diode V-I curves, the voltage across the diode changes only slightly due to changes in current drawn or changes in the input. When precise voltage control and efficiency are not important, this design may be fine.

Feedback voltage regulators operate by comparing the actual output voltage to some fixed reference voltage. Any difference is amplified and used to control the regulation element in such a way as to reduce the voltage error. This forms a negative feedback control loop; increasing the open-loop gain tends to increase regulation accuracy but reduce stability. (Stability is avoidance of oscillation, or ringing, during step changes.) There will also be a trade-off between stability and the speed of the response to changes. If the output voltage is too low (perhaps due to input voltage reducing or load current increasing), the regulation element is commanded, *up to a point*, to produce a higher output voltage–by dropping less of the input voltage (for linear series regulators and buck switching regulators), or to draw input current for longer periods (boost-type switching regulators); if the output voltage is too high, the regulation element will normally be commanded to produce a lower voltage. However, many regulators have over-current protection, so that they will entirely stop sourcing current (or limit the current in some way) if the output current is too high, and some regulators may also shut down if the input voltage is outside a given range.

Electromechanical Regulators

In electromechanical regulators, voltage regulation is easily accomplished by coiling the sensing wire to make an electromagnet. The magnetic field produced by the current attracts a moving ferrous core held back under spring tension or gravitational pull. As voltage increases, so does the current, strengthening the magnetic field produced by the coil and pulling the core towards the field. The magnet is physically connected to a mechanical power switch, which opens as the magnet moves into the field. As voltage decreases, so does the current, releasing spring tension or the weight of the core and causing it to retract. This closes the switch and allows the power to flow once more.

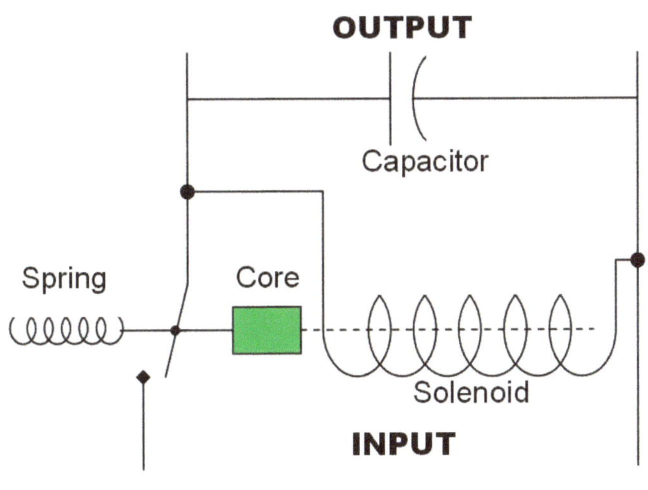

Circuit design for a simple electromechanical voltage regulator.

A voltage stabilizer using electromechanical relays for switching.

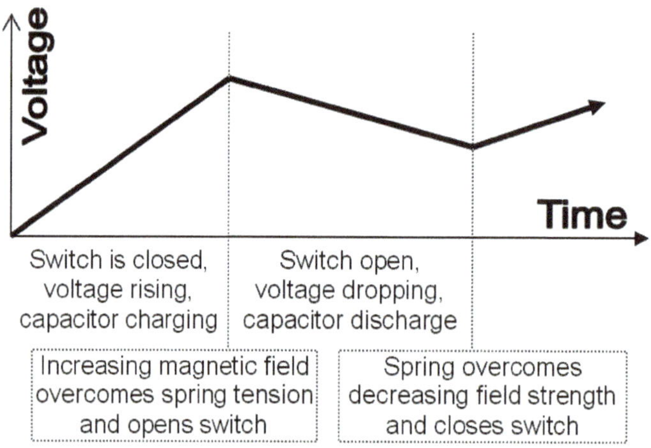

Graph of voltage output on a time scale.

If the mechanical regulator design is sensitive to small voltage fluctuations, the motion of the solenoid core can be used to move a selector switch across a range of resistances or transformer windings to gradually step the output voltage up or down, or to rotate the position of a moving-coil AC regulator.

Early automobile generators and alternators had a mechanical voltage regulator using one, two, or three relays and various resistors to stabilize the generator's output at slightly more than 6 or 12 V, independent of the engine's rpm or the varying load on the vehicle's electrical system. Essentially, the relay(s) employed pulse width modulation to regulate the output of the generator, controlling the field current reaching the generator (or alternator) and in this way controlling the output voltage producing back into the generator and attempting to run it as a motor. The rectifier diodes in an alternator automatically perform this function so that a specific relay is not required; this appreciably simplified the regulator design.

More modern designs now use *solid state* technology (transistors) to perform the same function that the relays perform in electromechanical regulators.

Electromechanical regulators are used for mains voltage stabilization — see AC voltage stabilizers below.

Automatic Voltage Regulator

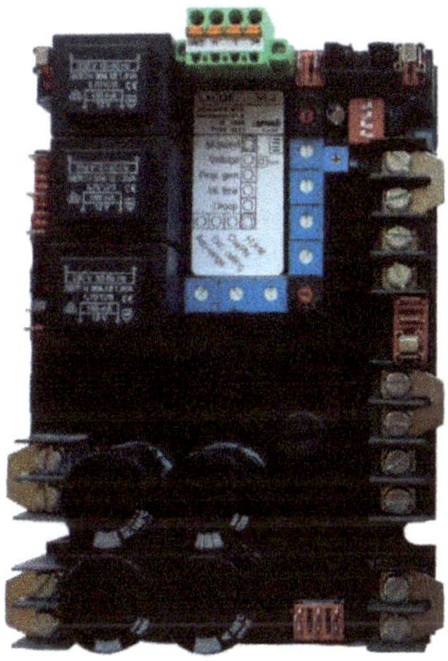

Voltage regulator for generators.

To control the output of generators (as seen in ships and power stations, or on oil rigs, greenhouses and emergency power systems) automatic voltage regulators are used. This is an active system. While the basic principle is the same, the system itself is more complex. An automatic voltage regulator (or AVR for short) consists of several components such as diodes, capacitors, resistors and potentiometers or even microcontrollers, all placed on a circuit board. This is then mounted near the generator and connected with several wires to measure and adjust the generator.

How an AVR works: In the first place the AVR monitors the output voltage and controls the input voltage for the exciter of the generator. By increasing or decreasing the generator control voltage, the output voltage of the generator increases or decreases accordingly. The AVR calculates how much voltage has to be sent to the exciter numerous times a second, therefore stabilizing the output voltage to a predetermined setpoint. When two or more generators are powering the same system (parallel operation) the AVR receives information from more generators to match all output.

AC Voltage Stabilizers

Coil-Rotation AC Voltage Regulator

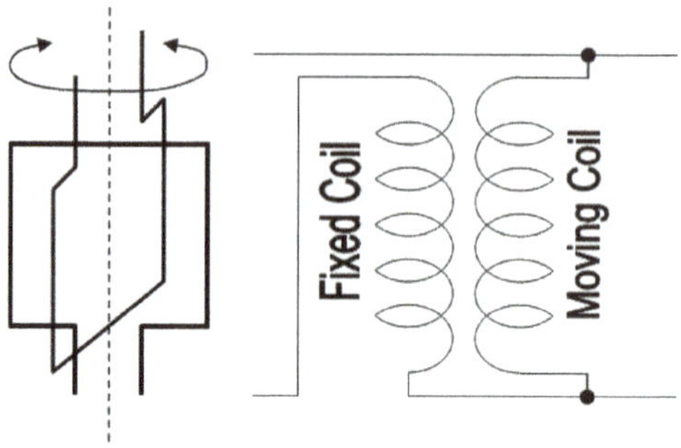

Basic design principle and circuit diagram for the rotating-coil AC voltage regulator.

This is an older type of regulator used in the 1920s that uses the principle of a fixed-position field coil and a second field coil that can be rotated on an axis in parallel with the fixed coil, similar to a variocoupler.

When the movable coil is positioned perpendicular to the fixed coil, the magnetic forces acting on the movable coil balance each other out and voltage output is unchanged. Rotating the coil in one direction or the other away from the center position will increase or decrease voltage in the secondary movable coil.

This type of regulator can be automated via a servo control mechanism to advance the movable coil position in order to provide voltage increase or decrease. A braking mechanism or high ratio gearing is used to hold the rotating coil in place against the powerful magnetic forces acting on the moving coil.

Magnetic mains regulator.

Electromechanical

Electromechanical regulators called *voltage stabilizers* or *tap-changers*, have also been used to regulate the voltage on AC power distribution lines. These regulators operate by using a servomechanism to select the appropriate tap on an autotransformer with multiple taps, or by moving the wiper on a continuously variable auto transfomer. If the output voltage is not in the acceptable range, the servomechanism switches the tap, changing the turns ratio of the transformer, to move the secondary voltage into the acceptable region. The controls provide a dead band wherein the controller will not act, preventing the controller from constantly adjusting the voltage ("hunting") as it varies by an acceptably small amount.

PWM Static Voltage Regulator

This is the latest technology of voltage regulation to provide real-time control of voltage fluctuation, sag, surge and also to control other power quality issues such as spikes and EMI/RFI electrical noises. This uses an IGBT regulator engine generating pulse width modulated (PWM) AC voltage at high switching frequency. This AC PWM wave is superimposed on the main incoming wave through a buck-boost transformer, to provide precisely regulated AC voltage. The regulation in this technology is instantaneous, thus making it suitable for electronic machines which need precise regulated power.

Constant-Voltage Transformer

The ferroresonant transformer, ferroresonant regulator or constant-voltage transformer is a type of saturating transformer used as a voltage regulator. These transformers use a tank circuit composed of a high-voltage resonant winding and a capacitor to produce a nearly constant average output voltage with a varying input current or varying load. The circuit has a primary on one side of a magnet shunt and the tuned circuit coil and secondary on the other side. The regulation is due to magnetic saturation in the section around the secondary.

The ferroresonant approach is attractive due to its lack of active components, relying on the square loop saturation characteristics of the tank circuit to absorb variations in average input voltage. Saturating transformers provide a simple rugged method to stabilize an AC power supply.

Older designs of ferroresonant transformers had an output with high harmonic content, leading to a distorted output waveform. Modern devices are used to construct a perfect sine wave. The ferroresonant action is a flux limiter rather than a voltage regulator, but with a fixed supply frequency it can maintain an almost constant average output voltage even as the input voltage varies widely.

The ferroresonant transformers, which are also known as Constant Voltage Transformers (CVTs) or ferros, are also good surge suppressors, as they provide high isolation and inherent short-circuit protection.

A ferroresonant transformer can operate with an input voltage range ±40% or more of the nominal voltage.

Output power factor remains in the range of 0.96 or higher from half to full load.

Because it regenerates an output voltage waveform, output distortion, which is typically less than 4%, is independent of any input voltage distortion, including notching.

Efficiency at full load is typically in the range of 89% to 93%. However, at low loads, efficiency can drop below 60%. The current-limiting capability also becomes a handicap when a CVT is used in an application with moderate to high inrush current like motors, transformers or magnets. In this case, the CVT has to be sized to accommodate the peak current, thus forcing it to run at low loads and poor efficiency.

Minimum maintenance is required, as transformers and capacitors can be very reliable. Some units have included redundant capacitors to allow several capacitors to fail between inspections without any noticeable effect on the device›s performance.

Output voltage varies about 1.2% for every 1% change in supply frequency. For example, a 2 Hz change in generator frequency, which is very large, results in an output voltage change of only 4%, which has little effect for most loads.

It accepts 100% single-phase switch-mode power supply loading without any requirement for derating, including all neutral components.

Input current distortion remains less than 8% THD even when supplying nonlinear loads with more than 100% current THD.

Drawbacks of CVTs are their larger size, audible humming sound, and the high heat generation caused by saturation.

Commercial Use

Voltage regulators or stabilizers are used to compensate for voltage fluctuations in mains power. Large regulators may be permanently installed on distribution lines. Small portable regulators may be plugged in between sensitive equipment and a wall outlet. Automatic voltage regulators are used on generator sets on ships, in emergency power supplies, on oil rigs, etc. to stabilize fluctuations in power demand. For example, when a large machine is turned on, the demand for power is suddenly a lot higher. The voltage regulator compensates for the change in load. Commercial voltage regulators normally operate on a range of voltages, for example 150–240 V or 90–280 V. Servo stabilizers are also manufactured and used widely in spite of the fact that they are obsolete and use out-dated technology.

Voltage regulators are used in devices like air conditioners, refrigerators, televisions etc. in order to protect them from fluctuating input voltage. The major problem faced is the use of relays in voltage regulators. Relays create sparks which result in faults in the product.

A three-phase bank of voltage regulators used to control the voltage on long AC power distribution lines. This bank is mounted on a wooden pole structure. Each regulator weighs about 2600 lbs and is rated 576 kVA.

DC Voltage Stabilizers

Many simple DC power supplies regulate the voltage using either series or shunt regulators, but most apply a voltage reference using a *shunt regulator* such as a Zener diode, avalanche breakdown diode, or voltage regulator tube. Each of these devices begins conducting at a specified voltage and will conduct as much current as required to hold its terminal voltage to that specified voltage by diverting excess current from a non-ideal power source to ground, often through a relatively low-value resistor to dissipate the excess energy. The power supply is designed to only supply a maximum amount of current that is within the safe operating capability of the shunt regulating device.

If the stabilizer must provide more power, the shunt regulator output is only used to provide the standard voltage reference for the electronic device, known as the voltage stabilizer. The voltage stabilizer is the electronic device, able to deliver much larger currents on demand.

Active Regulators

Active regulators employ at least one active (amplifying) component such as a transistor or operational amplifier. Shunt regulators are often (but not always) passive and simple, but always inefficient because they (essentially) dump the excess current not the load. When more power must be supplied, more sophisticated circuits are used. In general, these active regulators can be divided into several classes:

- Linear series regulators

- Switching regulators
- SCR regulators

Linear Regulators

Linear regulators are based on devices that operate in their linear region (in contrast, a switching regulator is based on a device forced to act as an on/off switch). Linear regulators are also classified in two types- 1. Series Regulators 2. Shunt Regulators

In the past, one or more vacuum tubes were commonly used as the variable resistance. Modern designs use one or more transistors instead, perhaps within an integrated circuit. Linear designs have the advantage of very "clean" output with little noise introduced into their DC output, but are most often much less efficient and unable to step-up or invert the input voltage like switched supplies. All linear regulators require a higher input than the output. If the input voltage approaches the desired output voltage, the regulator will "drop out". The input to output voltage differential at which this occurs is known as the regulator's drop-out voltage. Low-dropout regulators (LDOs) allow an input voltage that can be much lower (i.e., they waste less energy than conventional linear regulators).

Entire linear regulators are available as integrated circuits. These chips come in either fixed or adjustable voltage types.

Switching Regulators

Switching regulator integrated circuit LM2676, 3A step-down converter.

Switching regulators rapidly switch a series device on and off. The duty cycle of the switch sets how much charge is transferred to the load. This is controlled by a similar feedback mechanism as in a linear regulator. Because the series element is either fully conducting, or switched off, it dissipates almost no power; this is what gives the switching design its efficiency. Switching regulators are also able to generate output voltages which are higher than the input, or of opposite polarity — something not possible with a linear design.

Like linear regulators, nearly complete switching regulators are also available as integrated circuits. Unlike linear regulators, these usually require an inductor that acts as the energy storage element.

Comparing Linear Versus Switching Regulators

The two types of regulators have their different advantages:

- Linear regulators are best when low output noise (and low RFI radiated noise) is required
- Linear regulators are best when a fast response to input and output disturbances is required
- At low power levels, linear regulators are cheaper and occupy less printed circuit board space
- Switching regulators are best when power efficiency is critical (such as in portable computers), *except* that linear regulators are more efficient in a small number of cases (such as a 5 V microprocessor often in "sleep" mode fed from a 6 V battery, *if* the complexity of the switching circuit and the junction capacitance charging current means a high quiescent current in the switching regulator)
- Switching regulators are required when the only power supply is a DC voltage, and a higher output voltage is required.
- At power levels above a few watts, switching regulators are cheaper (for example, the cost of removing heat generated is less)

SCR Regulators

Regulators powered from AC power circuits can use silicon controlled rectifiers (SCRs) as the series device. Whenever the output voltage is below the desired value, the SCR is triggered, allowing electricity to flow into the load until the AC mains voltage passes through zero (ending the half cycle). SCR regulators have the advantages of being both very efficient and very simple, but because they can not terminate an ongoing half cycle of conduction, they are not capable of very accurate voltage regulation in response to rapidly changing loads. An alternative is the SCR shunt regulator which uses the regulator output as a trigger, both series and shunt designs are noisy, but powerful, as the device has a low on resistance.

Combination or Hybrid Regulators

Many power supplies use more than one regulating method in series. For example, the output from a switching regulator can be further regulated by a linear regulator. The switching regulator accepts a wide range of input voltages and efficiently generates a (somewhat noisy) voltage slightly above the ultimately desired output. That is followed by a linear regulator that generates exactly the desired voltage and eliminates nearly all the noise generated by the switching regulator. Other designs may use an SCR regulator as the "pre-regulator", followed by another type of regulator. An efficient way of creating a variable-voltage, accurate output power supply is to combine a multi-tapped transformer with an adjustable linear post-regulator.

Example Linear Regulators

Transistor Regulator

In the simplest case a common collector amplifier also known as emitter follower is used with the base of the regulating transistor connected directly to the voltage reference:

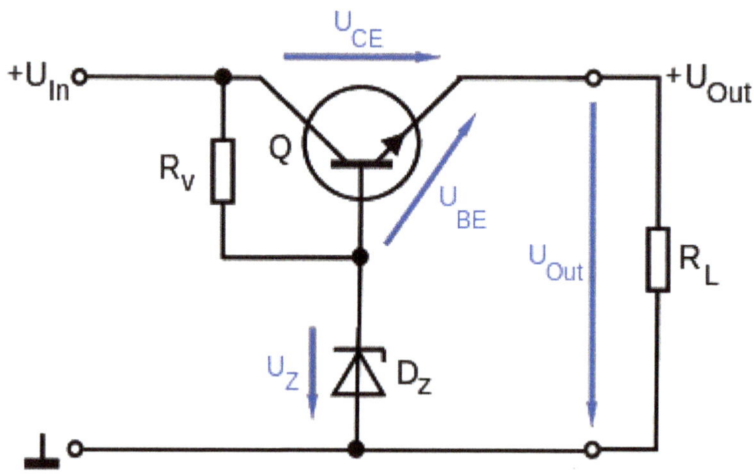

A simple transistor regulator will provide a relatively constant output voltage, U_{out}, for changes in the voltage of the power source, U_{in}, and for changes in load, R_L, provided that U_{in} exceeds U_{out} by a sufficient margin, and that the power handling capacity of the transistor is not exceeded.

The output voltage of the stabilizer is equal to the Zener diode voltage minus the base–emitter voltage of the transistor, $U_Z - U_{BE}$, where U_{BE} is usually about 0.7 V for a silicon transistor, depending on the load current. If the output voltage drops for any external reason, such as an increase in the current drawn by the load (causing a decrease in the Collector-Emitter junction voltage to observe KVL), the transistor's base–emitter voltage (U_{BE}) increases, turning the transistor on further and delivering more current to increase the load voltage again.

R_v provides a bias current for both the Zener diode and the transistor. The current in the diode is minimum when the load current is maximum. The circuit designer must choose a minimum voltage that can be tolerated across R_v, bearing in mind that the higher this voltage requirement is, the higher the required input voltage, U_{in}, and hence the lower the efficiency of the regulator. On the other hand, lower values of R_v lead to higher power dissipation in the diode and to inferior regulator characteristics.

R_v is given by:

$$R_v = \frac{V_{Rmin}}{I_{Dmin} + I_{Lmax}/(h_{FE} + 1)}$$

where $V_{R\,min}$ is the minimum voltage to be maintained across R_v

$I_{D\,min}$ is the minimum current to be maintained through the Zener diode

$I_{L\,max}$ is the maximum design load current

h_{FE} is the forward current gain of the transistor (I_C/I_B).

Regulator with an Operational Amplifier

The stability of the output voltage can be significantly increased by using an operational amplifier:

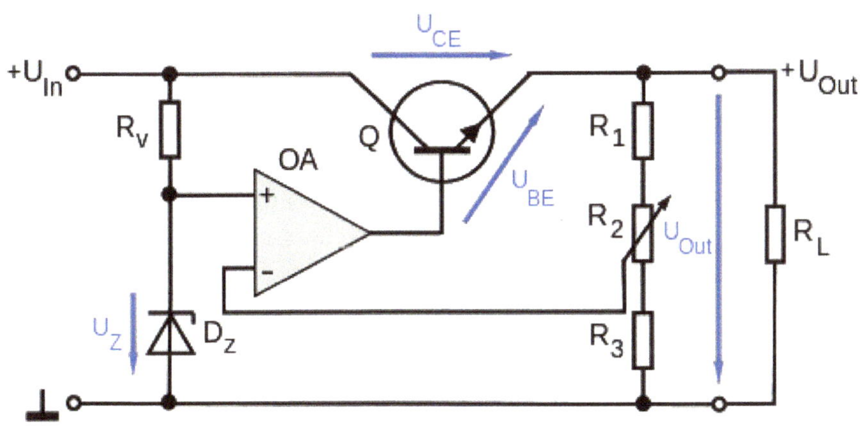

In this case, the operational amplifier drives the transistor with more current if the voltage at its inverting input drops below the output of the voltage reference at the non-inverting input. Using the voltage divider (R1, R2 and R3) allows choice of the arbitrary output voltage between U_z and U_{in}.

Measures of Regulator Quality

The output voltage can only be held *roughly* constant. The regulation is specified by two measurements:

- Load regulation is the change in output voltage for a given change in load current (for example, "typically 15 mV, maximum 100 mV for load currents between 5 mA and 1.4 A, at some specified temperature and input voltage").
- Line regulation or input regulation is the degree to which output voltage changes with input (supply) voltage changes - as a ratio of output to input change (for example, "typically 13 mV/V"), or the output voltage change over the entire specified input voltage range (for example, "plus or minus 2% for input voltages between 90 V and 260 V, 50-60 Hz").

Other important parameters are:

- Temperature coefficient of the output voltage is the change with temperature (perhaps averaged over a given temperature range).
- Initial accuracy of a voltage regulator (or simply "the voltage accuracy") reflects the error in output voltage for a fixed regulator without taking into account temperature or aging effects on output accuracy.
- Dropout voltage is the minimum difference between input voltage and output voltage for which the regulator can still supply the specified current. A low drop-out (LDO) regulator is designed to work well even with an input supply of only a Volt or so above the output voltage. The input-output differential at which the voltage regulator will no longer maintain regulation is the dropout voltage. Further reduction in input voltage will result in reduced output voltage. This value is dependent on load current and junction temperature.
- Absolute maximum ratings are defined for regulator components, specifying the continuous and peak output currents that may be used (sometimes internally limited), the maximum input voltage, maximum power dissipation at a given temperature, etc.
- Output noise (thermal white noise) and output dynamic impedance may be specified as graphs versus frequency, while output ripple noise (mains "hum" or switch-mode "hash" noise) may be given as peak-to-peak or RMS voltages, or in terms of their spectra.
- Quiescent current in a regulator circuit is the current drawn internally, not available to the load, normally measured as the input current while no load is connected and hence a source of inefficiency (some linear regulators are, surprisingly, more efficient at very low current loads than switch-mode designs because of this).
- Transient response is the reaction of a regulator when a (sudden) change of the load current (called the *load transient*) or input voltage (called the *line transient*) occurs. Some regulators will tend to oscillate or have a slow response time which in some cases might lead to undesired results. This value is different from the regulation parameters, as that is the stable situation definition. The

transient response shows the behaviour of the regulator on a change. This data is usually provided in the technical documentation of a regulator and is also dependent on output capacitance.

- Mirror-image insertion protection means that a regulator is designed for use when a voltage, usually not higher than the maximum input voltage of the regulator, is applied to its output pin while its input terminal is at a low voltage, volt-free or grounded. Some regulators can continuously withstand this situation. Others might only manage it for a limited time such as 60 seconds (usually specified in the data sheet). For instance, this situation can occur when a three terminal regulator is incorrectly mounted on a PCB, with the output terminal connected to the unregulated DC input and the input connected to the load. Mirror-image insertion protection is also important when a regulator circuit is used in battery charging circuits, when external power fails or is not turned on and the output terminal remains at battery voltage.

CYCLOCONVERTER

A cycloconverter (CCV) or a cycloinverter converts a constant voltage, constant frequency AC waveform to another AC waveform of a lower frequency by synthesizing the output waveform from segments of the AC supply without an intermediate DC link (Dorf 1993, pp. 2241–2243 and Lander 1993, p. 181). There are two main types of CCVs, circulating current type or blocking mode type, most commercial high power products being of the blocking mode type.

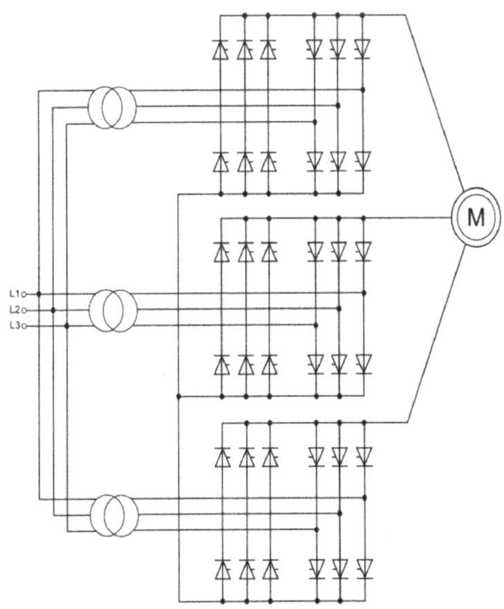

Topology of blocking mode cycloconverter.

Characteristics

Whereas phase-controlled SCR switching devices can be used throughout the range of CCVs, low cost, low-power TRIAC-based CCVs are inherently reserved for resistive load applications. The amplitude and frequency of converters' output voltage are both variable. The output to input frequency ratio of a three-phase CCV must be less than about one-third for circulating current mode CCVs or one-half for blocking mode CCVs.(Lander 1993, p. 188) Output waveform quality improves as the *pulse number* of switching-device bridges in phase-shifted configuration increases in CCV's input. In general, CCVs can be with 1-phase/1-phase, 3-phase/1-phase and 3-phase/3-phase input/output configurations, most applications however being 3-phase/3-phase.

Applications

The competitive power rating span of standardized CCVs ranges from few megawatts up to many tens of megawatts. CCVs are used for driving mine hoists, rolling mill main motors, ball mills for ore processing, cement kilns, ship propulsion systems, slip power recovery wound-rotor induction motors (i.e., Scherbius drives) and aircraft 400 Hz power generation. The variable-frequency output of a cycloconverter can be reduced essentially to zero. This means that very large motors can be started on full load at very slow revolutions, and brought gradually up to full speed. This is invaluable with, for example, ball mills, allowing starting with a full load rather than the alternative of having to start the mill with an empty barrel then progressively load it to full capacity. A fully loaded "hard start" for such equipment would essentially be applying full power to a stalled motor. Variable speed and reversing are essential to processes such as hot-rolling steel mills. Previously, SCR-controlled DC motors were used, needing regular brush/commutator servicing and delivering lower efficiency. Cycloconverter-driven synchronous motors need less maintenance and give greater reliability and efficiency. Single-phase bridge CCVs have also been used extensively in electric traction applications to for example produce 25 Hz power in the U.S. and 16 2/3 Hz power in Europe.

Whereas phase-controlled converters including CCVs are gradually being replaced by faster PWM self-controlled converters based on IGBT, GTO, IGCT and other switching devices, these older classical converters are still used at the higher end of the power rating range of these applications.

Harmonics

CCV operation creates current and voltage *harmonics* on the CCV's input and output. AC line harmonics are created on CCV's input accordance to the equation,

- $f_h = f_I\ (kq \pm 1) \pm 6nf_o$,

where

- f_h = harmonic frequency imposed on the AC line
- k and n = integers
- q = pulse number (6, 12 . . .)

- f_o = output frequency of the CCV
- Equation's 1st term represents the *pulse number* converter harmonic components starting with six-pulse configuration
- Equation's 2nd term denotes the converter's sideband characteristic frequencies including associated interharmonics and subharmonics.

VARIABLE-FREQUENCY TRANSFORMER

A variable-frequency transformer (VFT) is used to transmit electricity between two (asynchronous or synchronous) alternating current frequency domains. The VFT is a relatively recent development, first deployed in 2004. Most asynchronous grid inter-ties use high-voltage direct current converters, while synchronous grid inter-ties are connected by lines and "ordinary" transformers, but without the ability to control power flow between the systems.

Construction and Operation

A variable-frequency transformer is a doubly fed electric machine resembling a vertical shaft hydroelectric generator with a three-phase wound rotor, connected by slip rings to one external power circuit. The stator is connected to the other. With no applied torque, the shaft rotates due to the difference in frequency between the networks connected to the rotor and stator. A direct-current torque motor is mounted on the same shaft; changing the direction of torque applied to the shaft changes the direction of power flow.

The variable-frequency transformer behaves as a continuously adjustable phase-shifting transformer. It allows control of the power flow between two networks. Unlike power electronics solutions such as back-to-back HVDC, the variable frequency transformer does not demand harmonic filters and reactive power compensation. Limitations of the concept are the current-carrying capacity of the slip rings for the rotor winding.

Projects

Five small variable-frequency transformers with a total power rate of 25 MVA were in use at Neuhof Substation, Bad Sachsa, Germany for coupling power grids of former East and West Germany between 1985 and 1990.

Langlois Substation in Québec, Canada (45°17′13.76″N 74°0′56.07″W) installed a 100 MW variable-frequency transformer in 2004 to connect the asynchronous grids in Québec and the northeastern United States. This was the first large scale, commercial variable frequency transformer, and was installed at Hydro-Québec Langlois substation and is located electrically near sixteen hydro generators at Les Cèdres, Quebec and thirty-six more hydro generators at Beauharnois,_Quebec. The operating experience since April 2004 has demonstrated the VFT's inherent compatibility with the nearby generators

AEP Texas installed a 100 MW VFT substation in Laredo, Texas, United States (27°34′13.64″N 99°30′34.98″W) in early 2007. It connects the power systems of ERCOT (in the United States) to CFE (in Mexico).

Smaller VFTs are used in large land-based wind turbines, so that the turbine rotation speed can vary while connected to an electrical distribution grid.

Linden VFT

General Electric installed a 3 × 100 MW VFT substation in Linden, New Jersey, in the United States in 2009. It connects the power systems of PJM & New York Independent System Operator (NYISO). This installation is in parallel with three existing phase-shifting transformers to regulate synchronous power flow.

Economics of Energy Trading

VFTs provide the technical feasibility to flow power in both directions between two grids, permitting power exchanges that were previously impossible. Energy in a grid with lower costs can be transmitted to a grid with higher costs (higher demand), with energy trading. Power capacity is sold by providers. Transmission scheduling rights (TSRs) are auctioned by the transmission line owners.

Financial Transmission Rights (FTRs) are a financial instrument used to balance energy congestion and demand costs.

REFERENCES

1. J. W. Kolar, T. Friedli, F. Krismer, S. D. Round, "The Essence of Three-Phase AC/AC Converter Systems", Proceedings of the 13th Power Electronics and Motion Control Conference (EPE-PEMC'08), Poznan, Poland, pp. 27 – 42, Sept. 1 - 3, 2008.
2. Lee, M. Y. (2009). Three-level Neutral-point-clamped Matrix Converter Topology (PDF). University of Nottingham. p. 8.
3. I. Takahashi, Y. Itoh, "Electrolytic Capacitor-Less PWM Inverter", in Proceedings of the IPEC'90, Tokyo, Japan, , pp. 131 – 138, April 2 – 6, 1990.
4. K. Kuusela, M. Salo, H. Tuusa, "A Current Source PWM Converter Fed Permanent Magnet Synchronous Motor Drive with Adjustable DC-Link Current", in Proceedings of the NORPIE'2000, Aalborg, Denmark, pp. 54 – 58, June 15 – 16, 2000.
5. M. H. Bierhoff, F. W. Fuchs, "Pulse Width Modulation for Current Source Converters – A Detailed Concept," in Proceedings of the 32nd IEEE IECON'06, Paris, France, Nov. 7–10, 2006.
6. L. Gyugyi, B. R. Pelly, "Static Power Frequency Changers - Theory, Performance, & Application", New York: J. Wiley, 1976.
7. W. I. Popow, "Der zwangskommutierte Direktumrichter mit sinusförmiger Ausgangsspannung," Elektrie 28, No. 4, pp. 194 – 196, 1974
8. J. Holtz, U. Boelkens, "Direct Frequency Converter with Sinusoidal Line Currents for Speed-Variable AC Motors", IEEE Transactions Industry Electronics, Vol.

36, No. 4, pp. 475–479, 1989.

9. K. Shinohara, Y. Minari, T. Irisa, "Analysis and Fundamental Characteristics of Induction Motor Driven by Voltage Source Inverter without DC Link Components (in Japanese)", IEEJ Transactions, Vol. 109-D, No. 9, pp. 637 – 644, 1989.
10. L. Wei, T. A. Lipo, "A Novel Matrix Converter Topology with Simple Commutation", in Proceedings of the 36th IEEE IAS'01, Chicago, USA, vol. 3, pp. 1749–1754, Sept. 30 – Oct. 4, 2001.
11. Swamy, Mahesh; Kume, Tsuneo (Dec 16, 2010). "Present State and Futuristic Vision of Motor Drive Technology" (PDF). Power Transmission Engineering. www.powertransmission.com. Retrieved 8 October 2016.
12. Knowlton 1949, §6-128 Distribution Ttansformers, p. 597, Fig. 6–42
13. Mack, James E.; Shoemaker, Thomas (2006). Chapter 15 - Distribution Transformers (PDF) (11th ed.). New York: McGraw-Hill. pp. 15–1 to 15–22. ISBN 0-07-146789-0.
14. Bedell, Frederick. History of A-C Wave Form, Its Determination and Standardization. Transactions of the American Institute of Electrical Engineers. 61 (12): 864. doi:10.1109/T-AIEE.1942.5058456.
15. Brenner & Javid 1959, §18-1 Symbols and Polarity of Mutual Inductance, pp.=589–590
16. Brenner & Javid 1959, §18-6 The Ideal Transformer, pp.=598–600
17. Hameyer 2001, §2.1.2 Second Maxwell-Equation (Faraday›s Law) in Section 2 - Basics, pp. 11–12, eq. 2-12 to 2-15
18. Rajput, R.K. (2002). Alternating current s (3rd ed.). New Delhi: Laxmi Publications. p. 107. ISBN 9788170082224.
19. Hameyer 2001, §3.2 Definition of Transformer Ratio in Section 3 - Transformers, p. 27
20. Miller, Wilhelm C.; Robbins, Allan H. (2013). Circuit analysis : theory and practice (5th ed.). Clifton Park, NY: Cengage Learning. p. 990. ISBN 978-1-1332-8100-9. Retrieved 25 September 2014.
21. Flanagan 1993, pp. 1–2
22. Tcheslavski, Gleb V. (2008). Slide 13 Impedance Transformation in Lecture 4: Transformers. ELEN 3441 Fundamentals of Power Engineering. Lamar University (TSU system member).
23. Terman, Frederick E. (1955). Electronic and Radio Engineering (4th ed.). New York: McGraw-Hill. p. 15.
24. Knowlton 1949, §6-97 Nomenclature for Parallel Operation, pp. 585-586
25. IEC Std 60404-2 Magnetic Materials – Part 2: Methods of Measurement of the Magnetic Properties . . .
26. Parker, Ula & Webb 2005, 172, 1017; §2.5.5 Transformers & §10.1.3 The Ideal Transformer
27. Kothari & Nagrath 2010, p. 73, §3.7 Transformer Testing in Chapter 3 Transformers

28. Brenner & Javid 1959, §18-6 The Ideal Transformer, p.=589
29. Polarity Markings on Instrument Transformers (PDF). Retrieved 13 April 2013.
30. ANSI/IEEE C57.13, ANS Requirements for Instrument Transformers. New York, N.Y.: IEEE. 1978. p. 4 (§3.26). ISBN 0-7381-4299-9. (superseded, 1993)
31. Connections - Polarity (PDF). Retrieved 13 April 2013.
32. Hameyer 2001, in Section 2 Basics, p. 11, eq. 2-13
33. Billings, Keith (1999). Switchmode Power Supply Handbook. McGraw-Hill. ISBN 0-07-006719-8.
34. 400 Hz Electrical Systems. Aerospaceweb.org. Retrieved May 21, 2007.
35. Gururaj, B.I. (June 1963). Natural Frequencies of 3-Phase Transformer Windings. IEEE Transactions on Power Apparatus and Systems. 82 (66): 318–329. doi:10.1109/TPAS.1963.291359.
36. Kubo, T.; Sachs, H.; Nadel, S. (2001). Opportunities for New Appliance and Equipment Efficiency Standards. American Council for an Energy-Efficient Economy. p. 39, fig. 1. Retrieved June 21, 2009.
37. Riemersma, H.; Eckels, P.; Barton, M.; Murphy, J.; Litz, D.; Roach, J. (1981). Application of Superconducting Tcchnology to Powcr Transformers. IEEE Transactions on Power Apparatus and Systems. PAS-100 (7): 3398. doi:10.1109/TPAS.1981.316682.
38. Knowlton 1949, §2-67 Steinmetz› Formula; §4.279 Hysteris Loops, p.323
39. EE-Reviewonline.com. Steinmetz›s Formula for Magnetic Hysteresis. Retrieved 7 February 2013.
40. Nailen, Richard (May 2005). Why We Must Be Concerned With Transformers. Electrical Apparatus.
41. Hydroelectric Research and Technical Services Group. Transformers: Basics, Maintenance, and Diagnostics (PDF). U.S. Dept. of the Interior, Bureau of Reclamation. p. 12. Retrieved Mar 27, 2012.
42. Allan, D.J. (Jan 1991). Power Transformers – The Second Century. Power Engineering Journal. 5 (1): 5–14. doi:10.1049/pe:19910004.
43. Harlow 2004, §2.1.7 & §2.1.6.2.1 in Section §2.1 Power Transformers by H. Jin Sim and Scott H. Digby in Chapter 2 Equipment Types
44. Boteler, D. H.; Pirjola, R. J.; Nevanlinna, H. (1998). The Effects of Geomagnetic Disturbances On Electrical Systems at the Earth›s Surface. Advances in Space Research. 22: 17–27. doi:10.1016/S0273-1177(97)01096-X.
45. Hasegawa, Ryusuke (June 2, 2000). Present Status of Amorphous Soft Magnetic Alloys. Journal of Magnetism and Magnetic Materials. 215-216: 240–245. doi:10.1016/S0304-8853(00)00126-8.
46. McLyman 2004, p. 3-1
47. Toroidal Line Power Transformers. Power Ratings Tripled. | Magnetics Magazine. www.magneticsmagazine.com. Retrieved 2016-09-23.

48. Lee, Reuben. Air-Core Transformers. Electronic Transformers and Circuits. Retrieved May 22, 2007.
49. Dixon, Lloyd (2001). Power Transformer Design (PDF). Magnetics Design Handbook. Texas Instruments.
50. Lane, Keith (2007). The Basics of Large Dry-Type Transformers. EC&M. Retrieved 29 January 2013.
51. Heathcote 1998, pp. 720–723
52. Dixon, L.H., Jr. (1997). Eddy Current Losses in Transformer Windings (PDF). Texas Instrument: R2–1–to–R2–10.
53. Harlow 2004, §3.4.8 in Section 3.4 Load and Thermal Performance by Robert F. Tillman in Chapter 3 Ancillary Topics
54. Walling May 2007
55. Kimberly, E.E. Permissible Temperatures for Insulation. Retrieved 12 February 2013.
56. Pansini 1999, p. 32
57. Hartley, William H. (2003). Analysis of Transformer Failures. 36th Annual Conference of the International Association of Engineering Insurers. p. 7 (fig. 6). Retrieved 30 January 2013.
58. Hartley, William H. (~2011). An Analysis of Transformer Failures, Part 1 – 1988 through 1997. The Locomotive. Retrieved 30 January 2013.
59. Prevost, Thomas A.; et al. (Nov 2006). Estimation of Insulation Life Based on a Dual Temperature Aging Model (PDF). Weidmann. p. 1. Retrieved Mar 30, 2012.
60. Sen Feb 2011, PSERC Pub. 11-02
61. ASTDR ToxFAQs for Polychlorinated Biphenyls. 2001. Retrieved June 10, 2007.
62. AFBI (2011). 9. Contaminants (PDF). State of the Seas Report. Agri-Food and Biosciences Institute & Northern Ireland Environment Agency. p. 71. ISBN 978-1-907053-20-7. Unknown ID 9977.
63. McDonald, C. J.; Tourangeau, R. E. (1986). PCBs, Question and Answer Guide Concerning Polychlorinated Biphenyls (PDF). Government of Canada: Environment Canada Department. ISBN 0-662-14595-X. Retrieved Nov 7, 2007.
64. Mehta, S.P.; Aversa, N.; Walker, M.S. (Jul 1997). Transforming Transformers [Superconducting windings] (PDF). IEEE Spectrum. 34 (7): 43–49. doi:10.1109/6.609815. Retrieved 14 November 2012.
65. Pansini 1999, pp. 66–67
66. Fink & Beatty 1978, pp. 10–38 through 10–40
67. Figueroa, Elisa; et al. (Jan–Feb 2009). Low Frequency Heating Field Dry-Out of a 750 MVA 500 kV Auto Transformer (PDF). Electricity Today. Retrieved Feb 28, 2012.
68. Ryan 2004, pp. 416–417
69. Lawhead, Larry; Hamilton, Randy; Horak, John (2006). Three Phase Transformer

Winding Configurations and Differential Relay Compensation (PDF). Georgia Tech 60th Protective Relay Conference. pp. 8–10. Retrieved Feb 23, 2012.

70. Brown, BIll. Section 6 Grounding Systems (PDF). Schneider. pp. 9–12. Retrieved 18 January 2013.

71. Beltz, Robert; Peacock, Ian; Vilcheck, William (2000). Application Considerations for High Resistance Ground Retrofits in Pulp and Paper Mills. Pulp and Paper Industry Technical Conference. pp. 33–40. doi:10.1109/PAPCON.2000.854186.

72. Knowlton 1949, §6-7 Classification of Transformers, pp. 549-550

73. Power Transformers On Triad Magnetics. catalog.triadmagnetics.com. Retrieved 2016-09-23.

74. IEEE PES TC (Fall 2011). Discussion of Class I & II Terminology (PDF). IEEE PES Transformer Committee. p. slide 6. Retrieved 27 January 2013.

75. Knowlton 1949, §12-341 Grounding Transformers, p. 1085, fig. 12-95

76. AEMO; et al. (2012). Joint Consultation Paper - Western Metropolitan Melbourne Transmission Connection and Subtransmission Capacity. p. 11. Archived from the original (PDF) on 4 January 2013.

77. Heathcote 1998, p. 1

78. How the Electricity Grid Works. Retrieved 2016-09-23.

79. Poyser, Arthur William (1892). Magnetism and Electricity: A Manual for Students in Advanced Classes. London and New York: Longmans, Green, & Co. p. 285, fig. 248.

80. A Brief History of Electromagnetism (PDF).

81. Electromagnetism. Smithsonian Institution Archives.

82. MacPherson, Ph.D., Ryan C. Joseph Henry: The Rise of an American scientist.

83. Chow, Tai L. (2006). Introduction to Electromagnetic Theory: A Modern Perspective. Sudbury, Mass.: Jones and Bartlett Publishers. p. 171. ISBN 0-7637-3827-1.

84. Faraday, Michael (1834). Experimental Researches on Electricity, 7th Series. Philosophical Transactions of the Royal Society. 124: 77–122. doi:10.1098/rstl.1834.0008.

85. Yablochkov 1876, FR Pat. 115793, p=248

86. Subject-Matter Index 1883, p. 248

87. Stanley Transformer. Los Alamos National Laboratory; University of Florida. Retrieved Jan 9, 2009.

88. De Fonveille, W. (Jan 22, 1880). Gas and Electricity in Paris. Nature. 21 (534): 283. Bibcode:1880Natur..21..282D. doi:10.1038/021282b0. Retrieved Jan 9, 2009.

89. Károly, Simonyi. The Faraday Law With a Magnetic Ohm›s Law. Természet Világa. Retrieved Mar 1, 2012.

90. Lucas, J.R. Historical Development of the Transformer (PDF). IEE Sri Lanka Centre. Retrieved Mar 1, 2012.
91. Jeszenszky, Sándor. Electrostatics and Electrodynamics at Pest University in the Mid-19th Century (PDF). University of Pavia. Retrieved Mar 3, 2012.
92. Hungarian Inventors and Their Inventions. Institute for Developing Alternative Energy in Latin America. Retrieved Mar 3, 2012.
93. Bláthy, Ottó Titusz. Budapest University of Technology and Economics, National Technical Information Centre and Library. Retrieved Feb 29, 2012.
94. Bláthy, Ottó Titusz (1860–1939). Hungarian Patent Office. Retrieved Jan 29, 2004.
95. Zipernowsky, Déri & Bláthy 1886, US Patent 352 105
96. Smil, Vaclav (2005). Creating the Twentieth Century: Technical Innovations of 1867—1914 and Their Lasting Impact. Oxford: Oxford University Press. p. 71. ISBN 978-0-19-803774-3.
97. Nagy, Árpád Zoltán (Oct 11, 1996). Lecture to Mark the 100th Anniversary of the Discovery of the Electron in 1897 (preliminary text). Budapest. Retrieved July 9, 2009.
98. Oxford English Dictionary (2nd ed.). Oxford University Press. 1989.
99. Hospitalier, Édouard (1882). The Modern Applications of Electricity. Translated by Julius Maier. New York: D. Appleton & Co. p. 103.
100. Ottó Bláthy, Miksa Déri, Károly Zipernowsky. IEC Techline. Retrieved Apr 16, 2010.
101. Skrabec, Quentin R. (2007). George Westinghouse: Gentle Genius. Algora Publishing. p. 102. ISBN 978-0-87586-508-9.
102. International Electrotechnical Commission. Otto Blathy, Miksa Déri, Károly Zipernowsky. IEC History. Retrieved May 17, 2007.
103. Westinghouse 1887, US Patent 366 362
104. Neidhöfer, Gerhard (2008). Michael von Dolivo-Dobrowolsky and Three-Phase: The Beginnings of Modern e Technology and Power Supply (in German). In collaboration with VDE History of Electrical Engineering Committee (2 ed.). Berlin: VDE-Verl. ISBN 978-3-8007-3115-2.
105. Uth, Robert (Dec 12, 2000). Tesla Coil. Tesla: Master of Lightning. PBS.org. Retrieved May 20, 2008.
106. Tesla 1891, US Patent 454 622

CHAPTER 7

Rectifiers

Chapter Contents

A rectifier is an electrical device composed of one or more diodes that converts alternating current (AC) to direct current (DC). A diode is like a one-way valve that allows an electrical current to flow in only one direction. This process is called rectification.

A rectifier can take the shape of several different physical forms such as solid-state diodes, vacuum tube diodes, mercury arc valves, silicon-controlled rectifiers and various other silicon-based semiconductor switches.

INTRODUCTION

A rectifier is an electrical device that converts alternating current (AC), which periodically reverses direction, to direct current (DC), which flows in only one direction. The process is known as rectification. Physically, rectifiers take a number of forms, including vacuum tube diodes, mercury-arc valves, copper and selenium oxide rectifiers, semiconductor diodes, silicon-controlled rectifiers and other silicon-based semiconductor switches. Historically, even synchronous electromechanical switches and motors have been used. Early radio receivers, called crystal radios, used a "cat's whisker" of fine wire pressing on a crystal of galena (lead sulfide) to serve as a point-contact rectifier or "crystal detector".

Rectifiers have many uses, but are often found serving as components of DC power supplies and high-voltage direct current power transmission systems. Rectification may serve in roles other than to generate direct current for use as a source of power. As noted, detectors of radio signals serve as rectifiers. In gas heating systems flame rectification is used to detect presence of a flame.

Because of the alternating nature of the input AC sine wave, the process of rectification alone produces a DC current that, though unidirectional, consists of pulses of current. Many applications of rectifiers, such as power supplies for radio, television and computer equipment, require a *steady* constant DC current (as would be produced by a battery). In these applications the output of the rectifier is smoothed by an electronic filter (usually a capacitor) to produce a steady current.

More complex circuitry that performs the opposite function, converting DC to AC, is called an inverter.

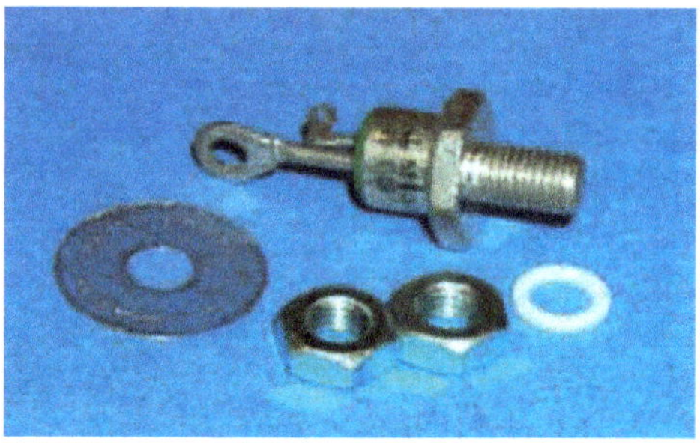

A rectifier diode (silicon controlled rectifier) and associated mounting hardware. The heavy threaded stud attaches the device to a heatsink to dissipate heat.

RECTIFIER DEVICES

Before the development of silicon semiconductor rectifiers, vacuum tube thermionic diodes and copper oxide- or selenium-based metal rectifier stacks were used. With the introduction of semiconductor electronics, vacuum tube rectifiers became obsolete, except for some enthusiasts of vacuum tube audio equipment. For power rectification from very low to very high current, semiconductor diodes of various types (junction diodes, Schottky diodes, etc.) are widely used.

Other devices that have control electrodes as well as acting as unidirectional current valves are used where more than simple rectification is required—e.g., where variable output voltage is needed. High-power rectifiers, such as those used in high-voltage direct current power transmission, employ silicon semiconductor devices of various types. These are thyristors or other controlled switching solid-state switches, which effectively function as diodes to pass current in only one direction.

RECTIFIER CIRCUITS

Rectifier circuits may be single-phase or multi-phase (three being the most common number of phases). Most low power rectifiers for domestic equipment are single-phase, but three-phase rectification is very important for industrial applications and for the transmission of energy as DC (HVDC).

Single-Phase Rectifiers

Half-Wave Rectification

In half-wave rectification of a single-phase supply, either the positive or negative half of the AC wave is passed, while the other half is blocked. Because only one half of the input waveform reaches the output, mean voltage is lower. Half-wave rectification requires a single diode in a single-phase supply, or three in a three-phase supply. Rectifiers yield a unidirectional but pulsating direct current; half-wave rectifiers produce far more ripple than full-wave rectifiers, and much more filtering is needed to eliminate harmonics of the AC frequency from the output.

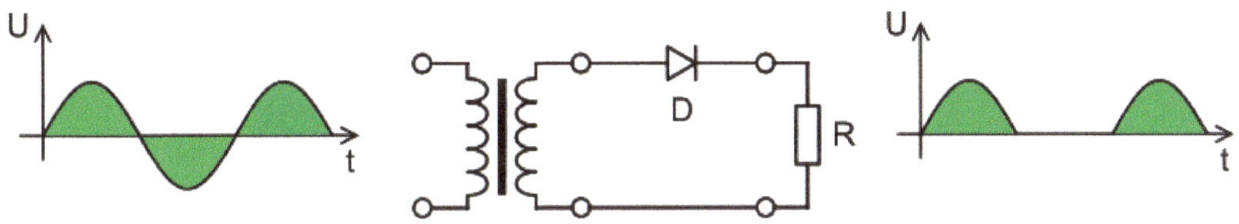

Half-wave rectifier.

The no-load output DC voltage of an ideal half-wave rectifier for a sinusoidal input voltage is:

$$V_{\mathrm{rms}} = \frac{V_{\mathrm{peak}}}{2}$$

$$V_{\mathrm{dc}} = \frac{V_{\mathrm{peak}}}{\pi}$$

where:

V_{dc}, V_{av} – the DC or average output voltage,

V_{peak}, the peak value of the phase input voltages,

V_{rms}, the root mean square (RMS) value of output voltage.

Full-Wave Rectification

A full-wave rectifier converts the whole of the input waveform to one of constant polarity (positive or negative) at its output. Full-wave rectification converts both polarities of the input waveform to pulsating DC (direct current), and yields a higher average output voltage. Two diodes and a center tapped transformer, or four diodes in a bridge configuration and any AC source (including a transformer without center tap), are needed. Single semiconductor diodes, double diodes with common cathode or common anode, and four-diode bridges, are manufactured as single components.

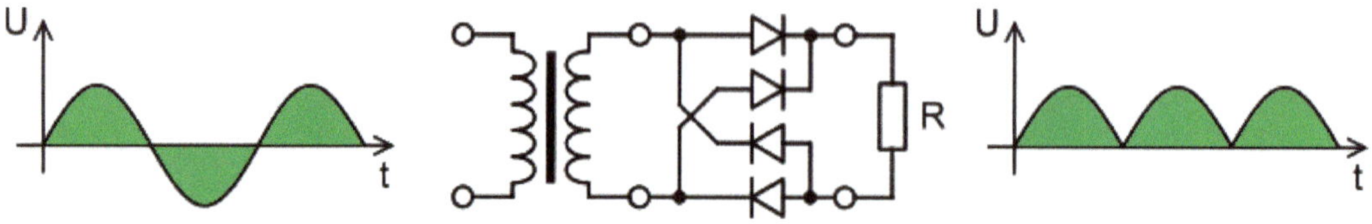

Graetz bridge rectifier: a full-wave rectifier using four diodes.

For single-phase AC, if the transformer is center-tapped, then two diodes back-to-back (cathode-to-cathode or anode-to-anode, depending upon output polarity required) can form a full-wave rectifier. Twice as many turns are required on the transformer secondary to obtain the same output voltage than for a bridge rectifier, but the power rating is unchanged.

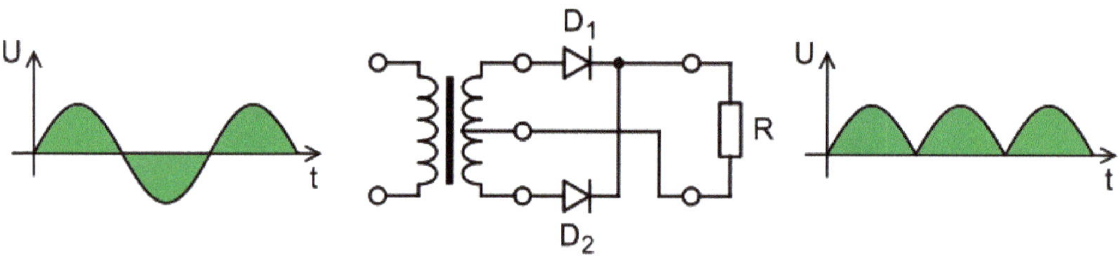

Full-wave rectifier using a center tap transformer and 2 diodes.

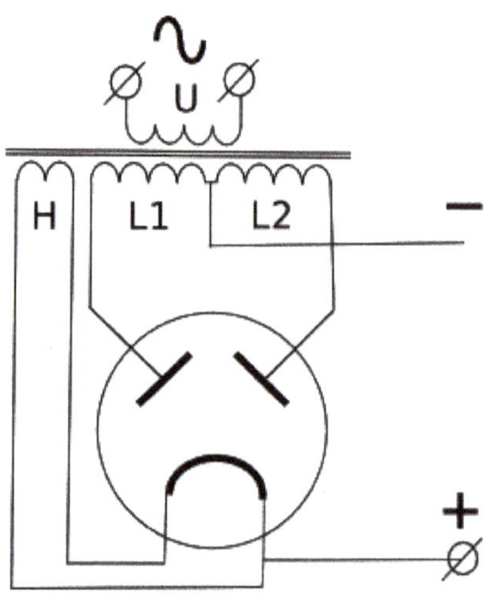

Full-wave rectifier, with vacuum tube having two anodes.

The average and RMS no-load output voltages of an ideal single-phase full-wave rectifier are:

$$V_{\mathrm{dc}} = V_{\mathrm{av}} = \frac{2V_{\mathrm{peak}}}{\pi}$$

$$V_{\mathrm{rms}} = \frac{V_{\mathrm{peak}}}{\sqrt{2}}$$

Very common double-diode rectifier vacuum tubes contained a single common cathode and two anodes inside a single envelope, achieving full-wave rectification with positive output. The 5U4 and 5Y3 were popular examples of this configuration.

Three-Phase Rectifiers

Single-phase rectifiers are commonly used for power supplies for domestic equipment. However, for most industrial and high-power applications, three-phase rectifier circuits are the norm. As with single-phase rectifiers, three-phase rectifiers can take the form of a half-wave circuit, a full-wave circuit using a center-tapped transformer, or a full-wave bridge circuit.

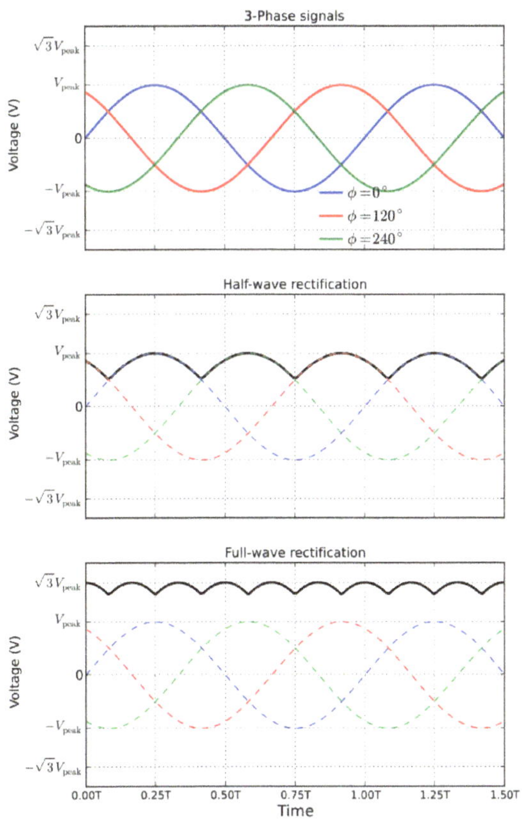

3-phase AC input, half- and full-wave rectified DC output waveforms.

Thyristors are commonly used in place of diodes to create a circuit that can regulate the output voltage. Many devices that provide direct current actually *generate* three-phase AC. For example, an automobile alternator contains six diodes, which function as a full-wave rectifier for battery charging.

Three-Phase, Half-Wave Circuit

An uncontrolled three-phase, half-wave circuit requires three diodes, one connected to each phase. This is the simplest type of three-phase rectifier but suffers from relatively high harmonic distortion on both the AC and DC connections. This type of rectifier is said to have a pulse-number of three, since the output voltage on the DC side contains three distinct pulses per cycle of the grid frequency.

Three-Phase, Full-Wave Circuit Using Center-Tapped Transformer

If the AC supply is fed via a transformer with a center tap, a rectifier circuit with improved harmonic performance can be obtained. This rectifier now requires six diodes, one connected to each end of each transformer secondary winding. This circuit has a pulse-number of six, and in effect, can be thought of as a six-phase, half-wave circuit.

Before solid state devices became available, the half-wave circuit, and the full-wave circuit using a center-tapped transformer, were very commonly used in industrial rectifiers using mercury-arc valves. This was because the three or six AC supply inputs could be fed to a corresponding number of anode electrodes on a single tank, sharing a common cathode.

With the advent of diodes and thyristors, these circuits have become less popular and the three-phase bridge circuit has become the most common circuit.

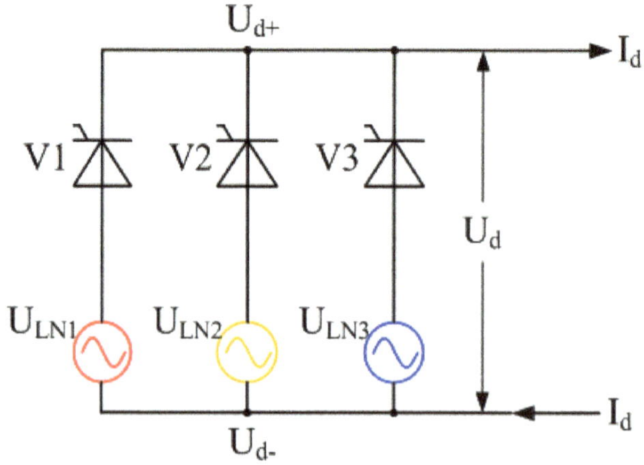

Three-phase half-wave rectifier circuit using thyristors as the switching elements, ignoring supply inductance.

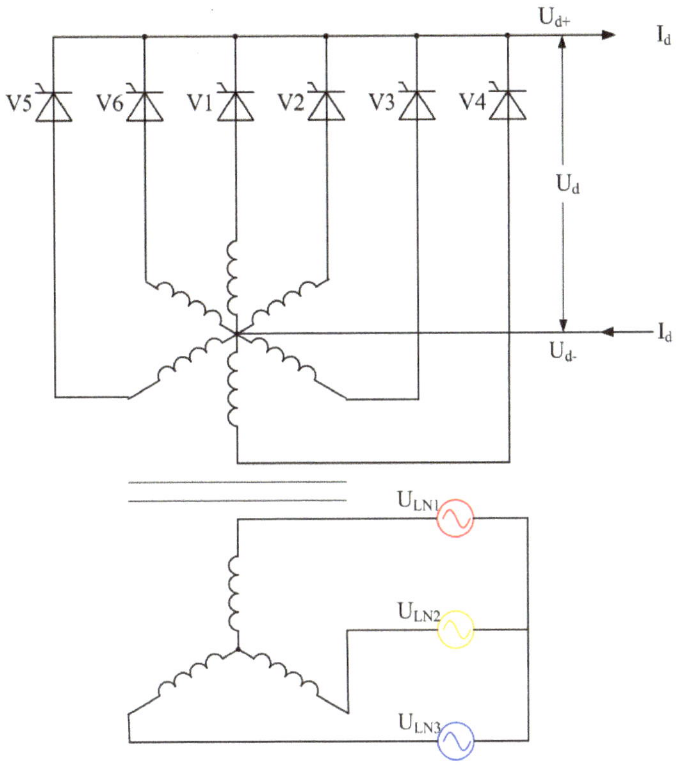

Three-phase full-wave rectifier circuit using thyristors as the switching elements, with a center-tapped transformer, ignoring supply inductance.

Three-Phase Bridge Rectifier

For an uncontrolled three-phase bridge rectifier, six diodes are used, and the circuit again has a pulse number of six. For this reason, it is also commonly referred to as a six-pulse bridge.

For low-power applications, double diodes in series, with the anode of the first diode connected to the cathode of the second, are manufactured as a single component for this purpose. Some commercially available double diodes have all four terminals available so the user can configure them for single-phase split supply use, half a bridge, or three-phase rectifier.

For higher-power applications, a single discrete device is usually used for each of the six arms of the bridge. For the very highest powers, each arm of the bridge may consist of tens or hundreds of separate devices in parallel (where very high current is needed, for example in aluminum smelting) or in series (where very high voltages are needed, for example in high-voltage direct current power transmission).

Disassembled automobile alternator, showing the six diodes that comprise a full-wave three-phase bridge rectifier.

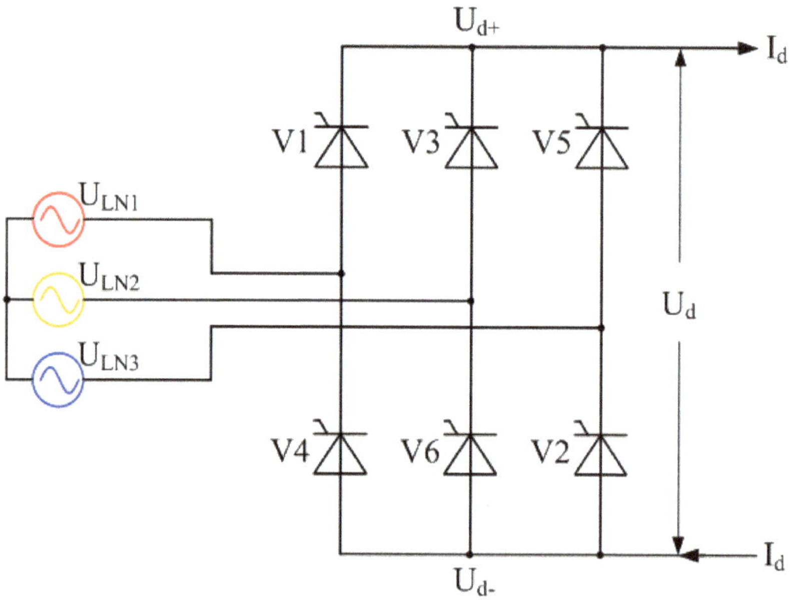

Three-phase full-wave bridge rectifier circuit using thyristors as the switching elements, ignoring supply inductance.

For a three-phase full-wave diode rectifier, the ideal, no-load average output voltage is

$$V_{\mathrm{dc}} = V_{\mathrm{av}} = \frac{3\sqrt{3}V_{\mathrm{peak}}}{\pi}$$

If thyristors are used in place of diodes, the output voltage is reduced by a factor cos(α):

$$V_{\mathrm{dc}} = V_{\mathrm{av}} = \frac{3\sqrt{3}V_{\mathrm{peak}}}{\pi}\cos\alpha$$

Or, expressed in terms of the line to line input voltage:

$$V_{\mathrm{dc}} = V_{\mathrm{av}} = \frac{3V_{\mathrm{LLpeak}}}{\pi}\cos\alpha$$

Where:

V_{LLpeak}, the peak value of the line to line input voltages,

V_{peak}, the peak value of the phase (line to neutral) input voltages,

α, firing angle of the thyristor (0 if diodes are used to perform rectification)

The above equations are only valid when no current is drawn from the AC supply or in the theoretical case when the AC supply connections have no inductance. In practice, the supply inductance causes a reduction of DC output voltage with increasing load, typically in the range 10–20% at full load.

The effect of supply inductance is to slow down the transfer process (called commutation) from one phase to the next. As result of this is that at each transition between a pair of devices, there is a period of overlap during which three (rather than two) devices in the bridge are conducting simultaneously. The overlap angle is usually referred to by the symbol μ (or u), and may be 20 30° at full load.

With supply inductance taken into account, the output voltage of the rectifier is reduced to:

$$V_{\mathrm{dc}} = V_{\mathrm{av}} = \frac{3V_{\mathrm{LLpeak}}}{\pi}\cos(\alpha+\mu)$$

The overlap angle μ is directly related to the DC current, and the above equation may be re-expressed as:

$$V_{\mathrm{dc}} = V_{\mathrm{av}} = \frac{3V_{\mathrm{LLpeak}}}{\pi}\cos(\alpha) - 6fL_{\mathrm{c}}I_{\mathrm{d}}$$

Where:

L_c, the commutating inductance per phase

I_d, the direct current

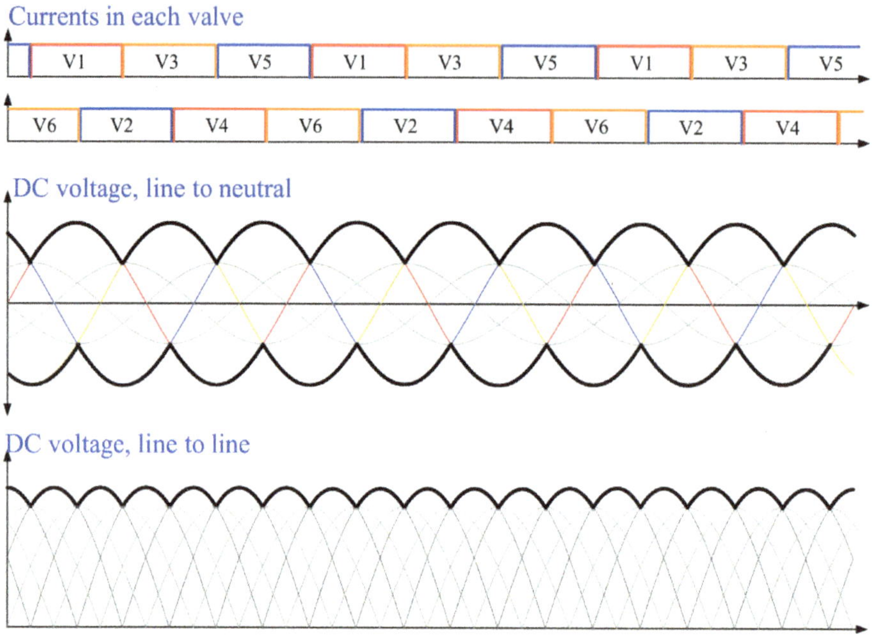

Three-phase Graetz bridge rectifier at alpha=0° without overlap.

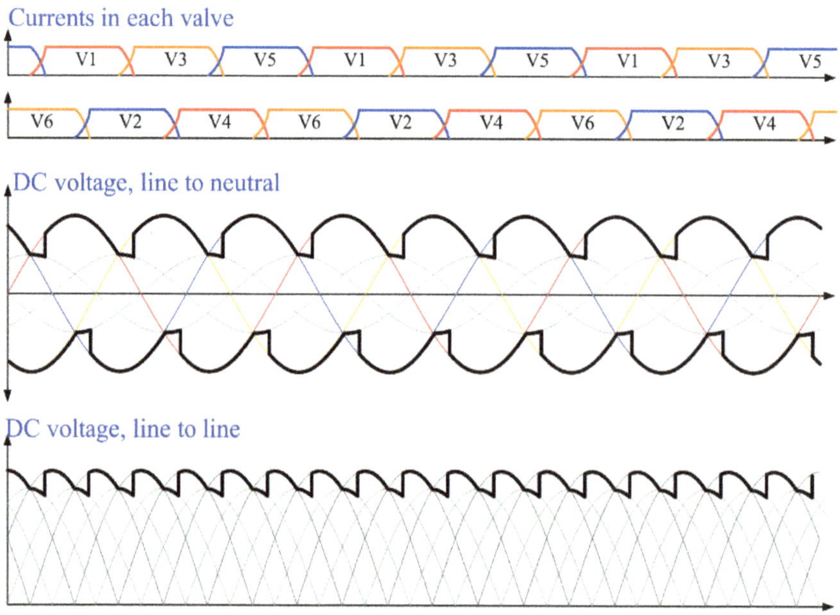

Three-phase Graetz bridge rectifier at alpha=0° with overlap angle of 20°.

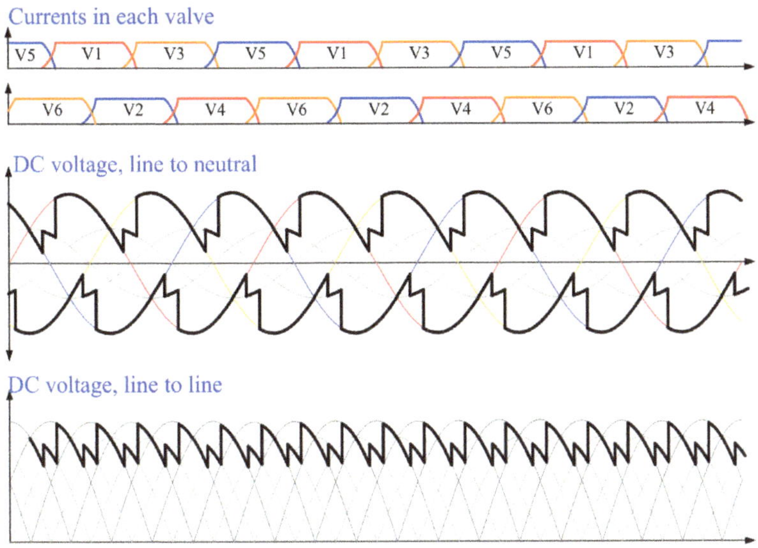

Three-phase controlled Graetz bridge rectifier at alpha=20° with overlap angle of 20°.

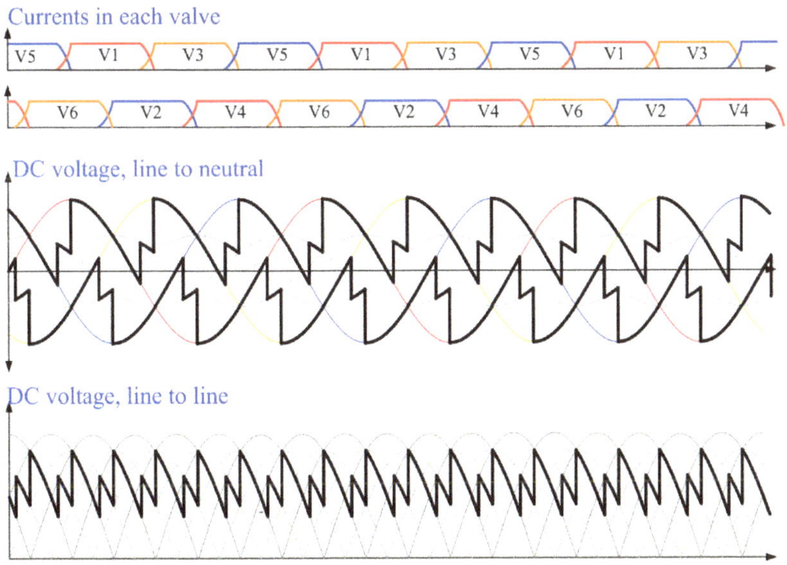

Three-phase controlled Graetz bridge rectifier at alpha=40° with overlap angle of 20°.

Twelve-Pulse Bridge

Although better than single-phase rectifiers or three-phase half-wave rectifiers, six-pulse rectifier circuits still produce considerable harmonic distortion on both the AC and DC connections. For very high-power rectifiers the twelve-pulse bridge connection is usually used. A twelve-pulse bridge consists of two six-pulse bridge circuits connected in series, with their AC connections fed from a supply transformer that produces a 30° phase shift between the two bridges. This cancels many of the characteristic harmonics the six-pulse bridges produce.

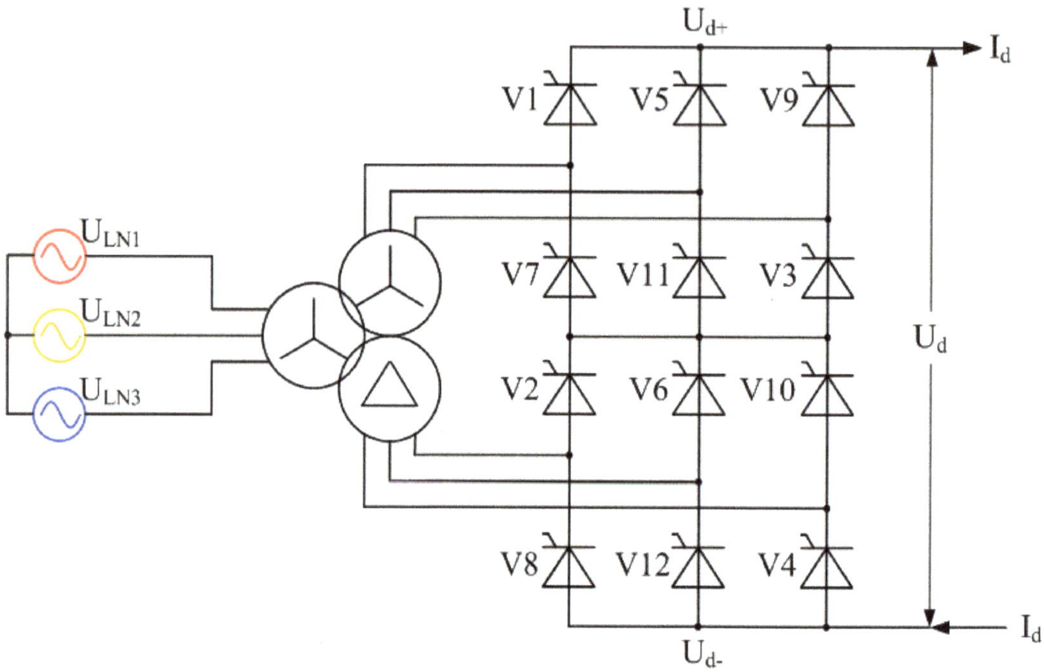

Twelve pulse bridge rectifier using thyristors as the switching elements.

The 30 degree phase shift is usually achieved by using a transformer with two sets of secondary windings, one in star (wye) connection and one in delta connection.

Voltage-Multiplying Rectifiers

The simple half-wave rectifier can be built in two electrical configurations with the diode pointing in opposite directions, one version connects the negative terminal of the output direct to the AC supply and the other connects the positive terminal of the output direct to the AC supply. By combining both of these with separate output smoothing it is possible to get an output voltage of nearly double the peak AC input voltage. This also provides a tap in the middle, which allows use of such a circuit as a split rail power supply.

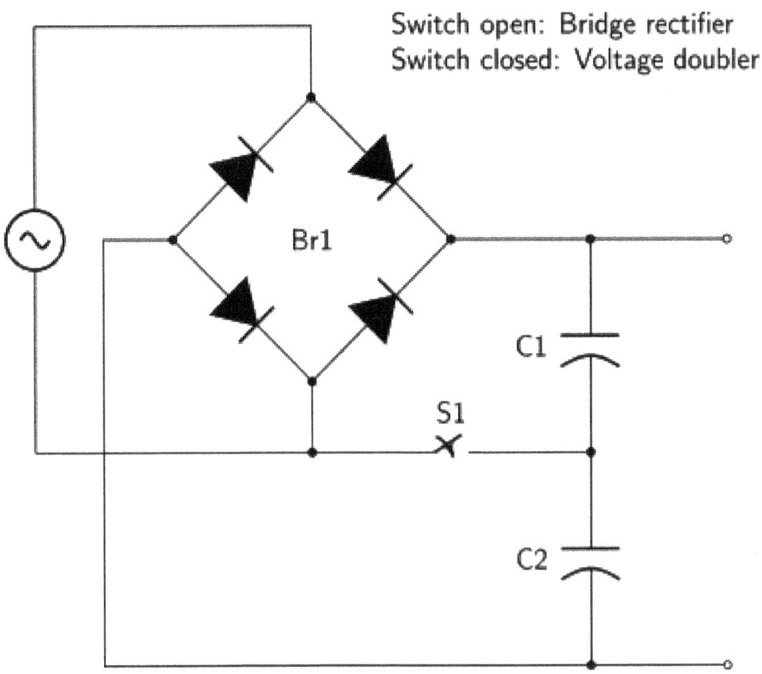

Switchable full bridge/voltage doubler.

A variant of this is to use two capacitors in series for the output smoothing on a bridge rectifier then place a switch between the midpoint of those capacitors and one of the AC input terminals. With the switch open, this circuit acts like a normal bridge rectifier. With the switch closed, it act like a voltage doubling rectifier. In other words, this makes it easy to derive a voltage of roughly 320 V (±15%, approx.) DC from any 120 V or 230 V mains supply in the world, this can then be fed into a relatively simple switched-mode power supply. However, for a given desired ripple, the value of both capacitors must be twice the value of the single one required for a normal bridge rectifier; when the switch is closed each one must filter the output of a half-wave rectifier, and when the switch is open the two capacitors are connected in series with an equivalent value of half one of them.

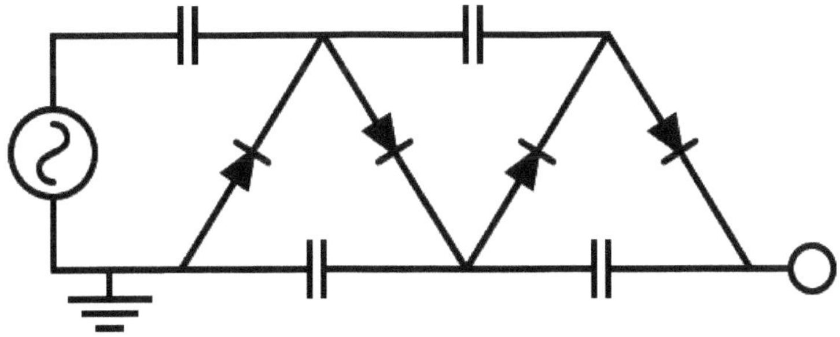

Cockcroft Walton voltage multiplier.

Cascaded diode and capacitor stages can be added to make a voltage multiplier (Cockroft-Walton circuit). These circuits are capable of producing a DC output voltage potential tens of times that of the peak AC input voltage, but are limited in current capacity and regulation. Diode voltage multipliers, frequently used as a trailing boost stage or primary high voltage (HV) source, are used in HV laser power supplies, powering devices such as cathode ray tubes (CRT) (like those used in CRT based television, radar and sonar displays), photon amplifying devices found in image intensifying and photo multiplier tubes (PMT), and magnetron based radio frequency (RF) devices used in radar transmitters and microwave ovens. Before the introduction of semiconductor electronics, transformerless powered vacuum tube receivers powered directly from AC power sometimes used voltage doublers to generate about 170 VDC from a 100–120 V power line.

RECTIFIER LOSSES

A real rectifier characteristically drops part of the input voltage (a voltage drop, for silicon devices, of typically 0.7 volts plus an equivalent resistance, in general non-linear)—and at high frequencies, distorts waveforms in other ways. Unlike an ideal rectifier, it dissipates some power. An aspect of most rectification is a loss from the peak input voltage to the peak output voltage, caused by the built-in voltage drop across the diodes (around 0.7 V for ordinary silicon p–n junction diodes and 0.3 V for Schottky diodes). Half-wave rectification and full-wave rectification using a center-tapped secondary produces a peak voltage loss of one diode drop. Bridge rectification has a loss of two diode drops. This reduces output voltage, and limits the available output voltage if a very low alternating voltage must be rectified. As the diodes do not conduct below this voltage, the circuit only passes current through for a portion of each half-cycle, causing short segments of zero voltage (where instantaneous input voltage is below one or two diode drops) to appear between each "hump". Peak loss is very important for low voltage rectifiers (for example, 12 V or less) but is insignificant in high-voltage applications such as HVDC.

RECTIFIER OUTPUT SMOOTHING

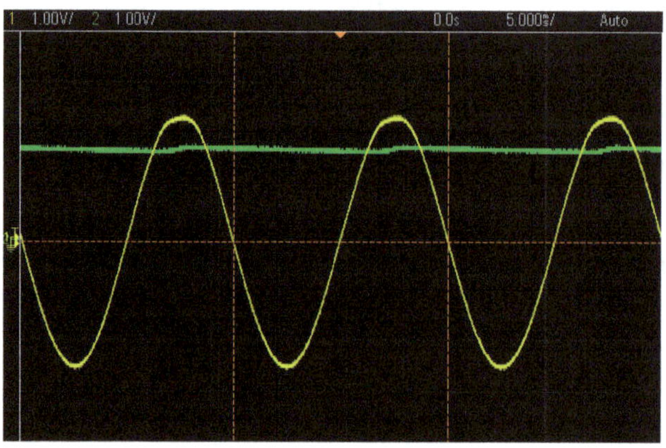

The AC input (yellow) and DC output (green) of a half-wave rectifier with a smoothing capacitor. Note the ripple in the DC signal.

While half-wave and full-wave rectification can deliver unidirectional current, neither produces a constant voltage. Producing steady DC from a rectified AC supply requires a smoothing circuit or filter. In its simplest form this can be just a reservoir capacitor or smoothing capacitor, placed at the DC output of the rectifier. There is still an AC ripple voltage component at the power supply frequency for a half-wave rectifier, twice that for full-wave, where the voltage is not completely smoothed.

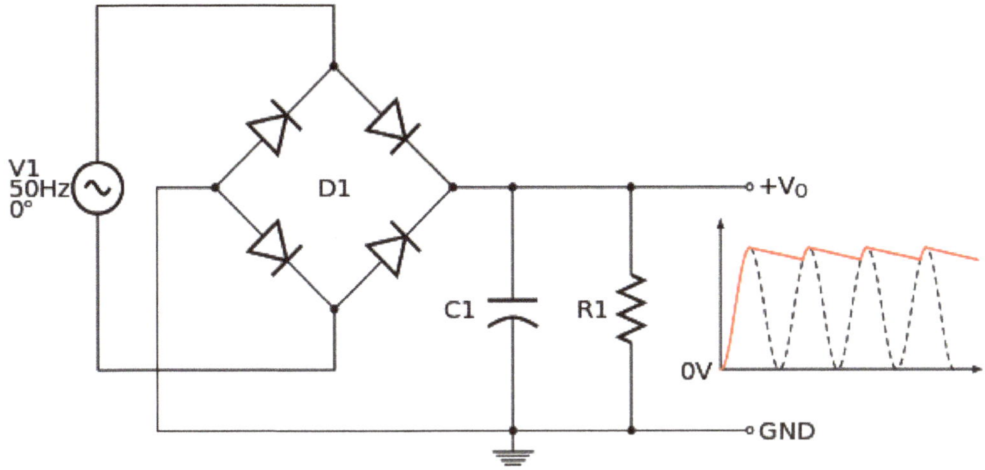

RC-Filter Rectifier: This circuit was designed and simulated using Multisim 8 software.

Sizing of the capacitor represents a tradeoff. For a given load, a larger capacitor reduces ripple but costs more and creates higher peak currents in the transformer secondary and in the supply that feeds it. The peak current is set in principle by the rate of rise of the supply voltage on the rising edge of the incoming sine-wave, but in practice it is reduced by the resistance of the transformer windings. In extreme cases where many rectifiers are loaded onto a power distribution circuit, peak currents may cause difficulty in maintaining a correctly shaped sinusoidal voltage on the ac supply.

To limit ripple to a specified value the required capacitor size is proportional to the load current and inversely proportional to the supply frequency and the number of output peaks of the rectifier per input cycle. The load current and the supply frequency are generally outside the control of the designer of the rectifier system but the number of peaks per input cycle can be affected by the choice of rectifier design.

A half-wave rectifier only gives one peak per cycle, and for this and other reasons is only used in very small power supplies. A full wave rectifier achieves two peaks per cycle, the best possible with a single-phase input. For three-phase inputs a three-phase bridge gives six peaks per cycle. Higher numbers of peaks can be achieved by using transformer networks placed before the rectifier to convert to a higher phase order.

To further reduce ripple, a capacitor-input filter can be used. This complements the reservoir capacitor with a choke (inductor) and a second filter capacitor, so that a steadier DC output can be obtained across the terminals of the filter capacitor. The choke presents a high impedance to the ripple current. For use at power-line frequencies inductors require cores of iron or other magnetic materials, and add weight and size. Their use in power supplies for electronic equipment has therefore dwindled in favor of semiconductor circuits such as voltage regulators.

A more usual alternative to a filter, and essential if the DC load requires very low ripple voltage, is to follow the reservoir capacitor with an active voltage regulator circuit. The reservoir capacitor must be large enough to prevent the troughs of the ripple dropping below the minimum voltage required by the regulator to produce the required output voltage. The regulator serves both to significantly reduce the ripple and to deal with variations in supply and load characteristics. It would be possible to use a smaller reservoir capacitor (these can be large on high-current power supplies) and then apply some filtering as well as the regulator, but this is not a common strategy. The extreme of this approach is to dispense with the reservoir capacitor altogether and put the rectified waveform straight into a choke-input filter. The advantage of this circuit is that the current waveform is smoother and consequently the rectifier no longer has to deal with the current as a large current pulse, but instead the current delivery is spread over the entire cycle. The disadvantage, apart from extra size and weight, is that the voltage output is much lower – approximately the average of an AC half-cycle rather than the peak.

APPLICATIONS

The primary application of rectifiers is to derive DC power from an AC supply (AC to DC converter). Virtually all electronic devices require DC, so rectifiers are used inside the power supplies of virtually all electronic equipment.

Converting DC power from one voltage to another is much more complicated. One method of DC-to-DC conversion first converts power to AC (using a device called an inverter), then uses a transformer to change the voltage, and finally rectifies power back to DC. A frequency of typically several tens of kilohertz is used, as this requires much smaller inductance than at lower frequencies and obviates the use of heavy, bulky, and expensive iron-cored units.

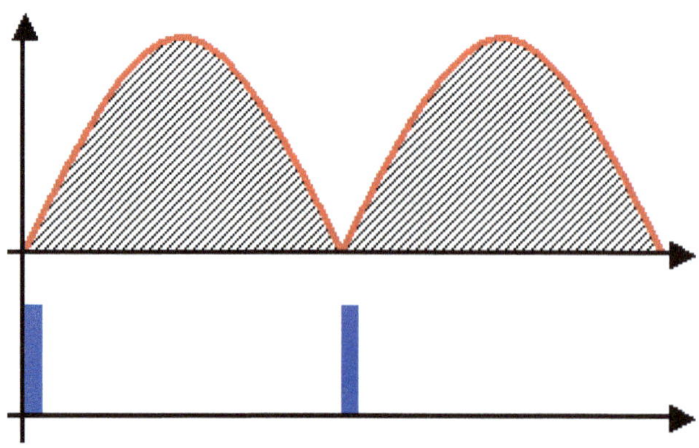

Output voltage of a full-wave rectifier with controlled thyristors.

Rectifiers are also used for detection of amplitude modulated radio signals. The signal may be amplified before detection. If not, a very low voltage drop diode or a diode biased with a fixed voltage must be used. When using a rectifier for demodulation the capacitor and load resistance must be carefully matched: too low a capacitance makes the high frequency carrier pass to the output, and too high makes the capacitor just charge and stay charged.

Rectifiers supply polarized voltage for welding. In such circuits control of the output current is required; this is sometimes achieved by replacing some of the diodes in a bridge rectifier with thyristors, effectively diodes whose voltage output can be regulated by switching on and off with phase fired controllers.

Thyristors are used in various classes of railway rolling stock systems so that fine control of the traction motors can be achieved. Gate turn-off thyristors are used to produce alternating current from a DC supply, for example on the Eurostar Trains to power the three-phase traction motors.

Rectification Teczhnologies

Electromechanical

Before about 1905 when tube type rectifiers were developed, power conversion devices were purely electro-mechanical in design. Mechanical rectification systems used some form of rotation or resonant vibration (e.g. vibrators) driven by electromagnets, which operated a switch or commutator to reverse the current.

These mechanical rectifiers were noisy and had high maintenance requirements. The moving parts had friction, which required lubrication and replacement due to wear. Opening mechanical contacts under load resulted in electrical arcs and sparks that heated and eroded the contacts. They also were not able to handle AC frequencies above several thousand cycles per second.

Synchronous Rectifier

To convert alternating into direct current in electric locomotives, a synchronous rectifier may be used. It consists of a synchronous motor driving a set of heavy-duty electrical contacts. The motor spins in time with the AC frequency and periodically reverses the connections to the load at an instant when the sinusoidal current goes through a zero-crossing. The contacts do not have to *switch* a large current, but they must be able to *carry* a large current to supply the locomotive's DC traction motors.

Vibrating Rectifier

These consisted of a resonant reed, vibrated by an alternating magnetic field created by an AC electromagnet, with contacts that reversed the direction of the current on the negative half cycles. They were used in low power devices, such as battery chargers, to rectify the low voltage produced by a step-down transformer. Another use was in battery power supplies for portable vacuum tube radios, to provide the high DC voltage for the tubes. These operated as a mechanical version of modern solid state switching inverters, with a transformer to step the battery voltage up, and a set of vibrator contacts on the transformer core, operated by its magnetic field, to repeatedly break the DC battery current to create a pulsing AC to power the transformer. Then a second set of rectifier contacts on the vibrator rectified the high AC voltage from the transformer secondary to DC.

A vibrator battery charger from 1922. It produced 6A DC at 6V to charge automobile batteries.

Motor-Generator Set

A small motor-generator set.

A *motor-generator set*, or the similar *rotary converter*, is not strictly a rectifier as it does not actually *rectify* current, but rather *generates* DC from an AC source. In an "M-G set", the shaft of an AC motor is mechanically coupled to that of a DC generator. The DC generator produces multiphase alternating currents in its armature windings, which a commutator on the armature shaft converts into a direct current output; or a homopolar generator produces a direct current without the need for a commutator. M-G sets are useful for producing DC for railway traction motors, industrial motors and other high-current applications, and were common in many high-power D.C. uses (for example, carbon-arc lamp projectors for outdoor theaters) before high-power semiconductors became widely available.

Electrolytic

The electrolytic rectifier was a device from the early twentieth century that is no longer used. A home-made version is illustrated in the 1913 book *The Boy Mechanic* but it would only be suitable for use at very low voltages because of the low breakdown voltage and the risk of electric shock. A more complex device of this kind was patented by G. W. Carpenter in 1928 (US Patent 1671970).

When two different metals are suspended in an electrolyte solution, direct current flowing one way through the solution sees less resistance than in the other direction. Electrolytic rectifiers most commonly used an aluminum anode and a lead or steel cathode, suspended in a solution of tri-ammonium ortho-phosphate.

The rectification action is due to a thin coating of aluminum hydroxide on the aluminum electrode, formed by first applying a strong current to the cell to build up the coating. The rectification process is temperature-sensitive, and for best efficiency should not operate above 86 °F (30 °C). There is also a breakdown voltage where the coating is penetrated and the cell is short-circuited. Electrochemical methods are often more fragile than mechanical methods, and can be sensitive to usage variations, which can drastically change or completely disrupt the rectification processes.

Similar electrolytic devices were used as lightning arresters around the same era by suspending many aluminium cones in a tank of tri-ammonium ortho-phosphate solution. Unlike the rectifier above, only aluminium electrodes were used, and used on A.C., there was no polarization and thus no rectifier action, but the chemistry was similar.

The modern electrolytic capacitor, an essential component of most rectifier circuit configurations was also developed from the electrolytic rectifier.

Plasma Type

The development of vacuum tube technology in the early 20th century resulted in the invention of various tube-type rectifiers, which largely replaced the noisy, inefficient mechanical rectifiers.

Mercury-Arc

Early 3-phase industrial mercury vapor rectifier tube.

150 kV mercury-arc valve at Manitoba Hydro power station, Radisson, Canada converted AC hydropower to DC for transmission to distant cities.

A rectifier used in high-voltage direct current (HVDC) power transmission systems and industrial processing between about 1909 to 1975 is a *mercury-arc rectifier* or *mercury-arc valve*. The device is enclosed in a bulbous glass vessel or large metal tub. One electrode, the cathode, is submerged in a pool of liquid mercury at the bottom of the vessel and one or more high purity graphite electrodes, called anodes, are suspended above the pool. There may be several auxiliary electrodes to aid in starting and maintaining the arc. When an electric arc is established between the cathode pool and suspended anodes, a stream of electrons flows from the cathode to the anodes through the ionized mercury, but not the other way (in principle, this is a higher-power counterpart to flame rectification, which uses the same one-way current transmission properties of the plasma naturally present in a flame).

These devices can be used at power levels of hundreds of kilowatts, and may be built to handle one to six phases of AC current. Mercury-arc rectifiers have been replaced by silicon semiconductor rectifiers and high-power thyristor circuits in the mid 1970s. The most powerful mercury-arc rectifiers ever built were installed in the Manitoba Hydro Nelson River Bipole HVDC project, with a combined rating of more than 1 GW and 450 kV.

Argon Gas Electron Tube

Tungar bulbs from 1917, 2 ampere *(left)* and 6 ampere.

The General Electric Tungar rectifier was an argon gas-filled electron tube device with a tungsten filament cathode and a carbon button anode. It operated similarly to the thermionic vacuum tube diode, but the gas in the tube ionized during forward conduction, giving it a much lower forward voltage drop so it could rectify lower voltages. It was used for battery chargers and similar applications from the 1920s until lower-cost metal rectifiers, and later semiconductor diodes, supplanted it. These were made up to a few hundred volts and a few amperes rating, and in some sizes strongly resembled an incandescent lamp with an additional electrode.

The 0Z4 was a gas-filled rectifier tube commonly used in vacuum tube car radios in the 1940s and 1950s. It was a conventional full-wave rectifier tube with two anodes and one cathode, but was unique in that it had no filament (thus the "0" in its type number). The electrodes were shaped such that the reverse breakdown voltage was much higher than the forward breakdown voltage. Once the breakdown voltage was exceeded, the 0Z4 switched to a low-resistance state with a forward voltage drop of about 24 V.

Diode Vacuum Tube (Valve)

The thermionic vacuum tube diode, originally called the Fleming valve, was invented by John Ambrose Fleming in 1904 as a detector for radio waves in radio receivers, and evolved into a general rectifier. It consisted of an evacuated glass bulb with a filament heated by a separate current, and a metal plate anode. The filament emitted electrons by thermionic emission (the Edison effect), discovered by Thomas Edison in 1884, and a positive voltage on the plate caused a current of electrons through the tube from filament to plate. Since only the filament produced electrons, the tube would only conduct current in one direction, allowing the tube to rectify an alternating current.

Vacuum tube diodes.

Vacuum diode rectifiers were widely used in power supplies in vacuum tube consumer electronic products, such as phonographs, radios, and televisions, for example the All American Five radio receiver, to provide the high DC plate voltage needed by other vacuum tubes. "Full-wave" versions with two separate plates were popular because they could be used with a center-tapped transformer to make a full-wave rectifier. Vacuum rectifiers were made for very high voltages, such as the high voltage power supply for the cathode ray tube of television receivers, and the kenotron used for power supply in X-ray equipment. However, compared to modern semiconductor diodes, vacuum rectifiers have high internal resistance due to space charge and therefore high voltage drops, causing high power dissipation and low efficiency. They are rarely able to handle currents exceeding 250 mA owing to the limits of plate power dissipation, and cannot be used for low voltage applications, such as battery chargers. Another limitation of the vacuum tube rectifier is that the heater power supply often requires special arrangements to insulate it from the high voltages of the rectifier circuit.

In musical instrument amplification (especially for electric guitars), the slight delay or "sag" between a signal increase (for instance, when a guitar chord is struck hard and fast) and the corresponding increase in output voltage is a notable effect of tube rectification, and results in compression. The choice between tube rectification and diode rectification is a matter of taste; some amplifiers have both and allow the player to choose.

Solid State

Crystal Detector

Galena cat's whisker detector.

The cat's-whisker detector was the earliest type of semiconductor diode. It consisted of a crystal of some semiconducting mineral, usually galena (lead sulfide), with a light springy wire touching its surface. Invented by Jagadish Chandra Bose and developed by G. W. Pickard around 1906, it served as the radio wave rectifier in the first widely used radio receivers, called crystal radios. Its fragility and limited current capability made it unsuitable for power supply applications. It became obsolete around 1920, but later versions served as microwave detectors and mixers in radar receivers during World War 2.

Selenium and Copper Oxide Rectifiers

Once common until replaced by more compact and less costly silicon solid-state rectifiers in the 1970s, these units used stacks of metal plates and took advantage of the semiconductor properties of selenium or copper oxide. While selenium rectifiers were lighter in weight and used less power than comparable vacuum tube rectifiers, they had the disadvantage of finite life expectancy, increasing resistance with age, and were only suitable to use at low frequencies. Both selenium and copper oxide rectifiers have somewhat better tolerance of momentary voltage transients than silicon rectifiers.

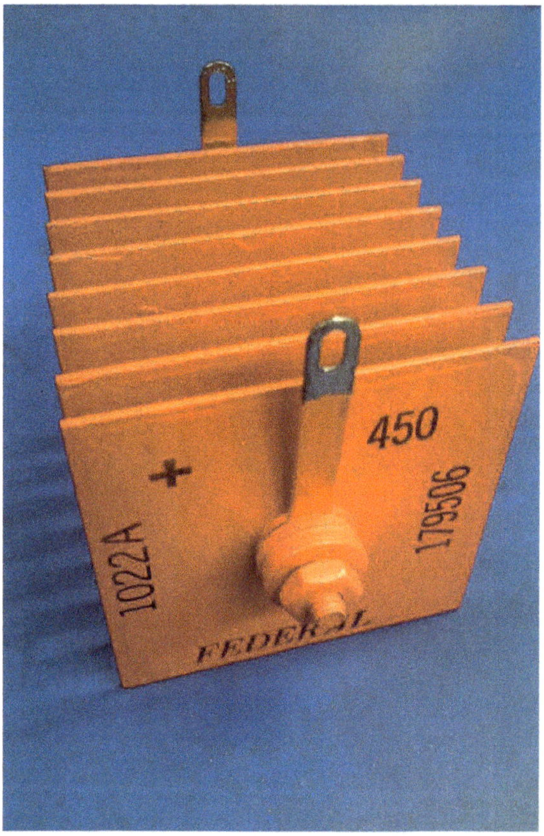

Selenium rectifier.

Typically these rectifiers were made up of stacks of metal plates or washers, held together by a central bolt, with the number of stacks determined by voltage; each cell was rated for about 20 V. An automotive battery charger rectifier might have only one cell: the high-voltage power supply for a vacuum tube might have dozens of stacked plates. Current density in an air-cooled selenium stack was about 600 mA per square inch of active area (about 90 mA per square centimeter).

Silicon and Germanium Diodes

In the modern world, silicon diodes are the most widely used rectifiers for lower voltages and powers, and have largely replaced earlier germanium diodes. For very high voltages and powers, the added need for controllability has in practice led to replacing simple silicon diodes with high-power thyristors (see below) and their newer actively gate-controlled cousins.

High power: Thyristors (SCRS) And Newer Silicon-Based Voltage Sourced Converters

Two of three high-power thyristor valve stacks used for long distance transmission of power from Manitoba Hydro dams. Compare with mercury-arc system from the same dam-site, above.

In high-power applications, from 1975 to 2000, most mercury valve arc-rectifiers were replaced by stacks of very high power thyristors, silicon devices with two extra layers of semiconductor, in comparison to a simple diode.

In medium-power transmission applications, even more complex and sophisticated voltage sourced converter (VSC) silicon semiconductor rectifier systems, such as insulated gate bipolar transistors (IGBT) and gate turn-off thyristors (GTO), have made smaller high voltage DC power transmission systems economical. All of these devices function as rectifiers.

As of 2009 it was expected that these high-power silicon “self-commutating switches”, in particular IGBTs and a variant thyristor (related to the GTO) called the integrated gate-commutated thyristor (IGCT), would be scaled-up in power rating to the point that they would eventually replace simple thyristor-based AC rectification systems for the highest power-transmission DC applications.

Current Research

A major area of research is to develop higher frequency rectifiers, that can rectify into terahertz and light frequencies. These devices are used in optical heterodyne detection, which has myriad applications in optical fiber communication and atomic clocks. Another prospective application for such devices is to directly rectify light waves picked up by tiny antenna, called nantennas, to produce DC electric power. It is thought that arrays of nantennas could be a more efficient means of producing solar power than solar cells.

A related area of research is to develop smaller rectifiers, because a smaller device has a higher cutoff frequency. Research projects are attempting to develop a unimolecular rectifier, a single organic molecule that would function as a rectifier.

REFERENCES

1. Morris, Peter Robin (1990). A History of the World Semiconductor Industry. p. 18. ISBN 978-0-86341-227-1.
2. Lander, Cyril W. (1993). "2. Rectifying Circuits". Power electronics (3rd ed.). London: McGraw-Hill. ISBN 978-0-07-707714-3.
3. Williams, B. W. (1992). "Chapter 11". Power electronics : devices, drivers and applications (2nd ed.). Basingstoke: Macmillan. ISBN 978-0-333-57351-8.
4. Hendrik Rissik (1941). Mercury-arc current convertors [sic] : an introduction to the theory and practice of vapour-arc discharge devices and to the study of rectification phenomena. Sir I. Pitman & sons, ltd. Retrieved 8 January 2013.
5. Kimbark, Edward Wilson (1971). Direct current transmission. (4. printing. ed.). New York: Wiley-Interscience. p. 508. ISBN 978-0-471-47580-4.
6. "Analog and Digital Electronics", Rectifier and clipper circuit, pages 4 and 5, retrieved 25 April 2015.
7. Wendy Middleton, Mac E. Van Valkenburg (eds), Reference Data for Engineers: Radio, Electronics, Computer, and Communications, p. 14. 13, Newnes, 2002 ISBN 0-7506-7291-9.
8. Mansell, A.D.; Shen, J. (1 January 1994). "Pulse converters in traction applications". Power Engineering Journal. 8 (4): 183. doi:10.1049/pe:19940407.
9. Hawkins, Nehemiah (1914). "54. Rectifiers". Hawkins Electrical Guide: Principles of electricity, magnetism, induction, experiments, dynamo. New York: T. Audel. Retrieved 8 January 2013.
10. "How To Make An Electrolytic Rectifier". Chestofbooks.com. Retrieved 2012-03-15.
11. US patent 1671970, Glenn W. Carpenter, "Liquid Rectifier", issued 1928-06-05
12. American Technical Society (1920). Cyclopedia of applied electricity. 2. American

technical society. p. 487. Retrieved 8 January 2013.

13. Pictures of a mercury-arc rectifier in operation can be seen here: Belsize Park deep shelter rectifier 1, Belsize Park deep shelter rectifier 2
14. Sood, Vijay K. HVDC and FACTS Controllers: Applications Of Static Converters In Power Systems. Springer-Verlag. p. 1. ISBN 978-1-4020-7890-3. The first 25 years of HVDC transmission were sustained by converters having mercury arc valves till the mid-1970s. The next 25 years till the year 2000 were sustained by line-commutated converters using thyristor valves. It is predicted that the next 25 years will be dominated by force-commutated converters . Initially, this new force-commutated era has commenced with Capacitor Commutated Converters (CCC) eventually to be replaced by self-commutated converters due to the economic availability of high-power switching devices with their superior characteristics.
15. Hunter, Dave (September 2013). "What's The Big Deal About Tube Rectification?". Guitar Player. p. 136.
16. H. P. Westman et al., (ed), Reference Data for Radio Engineers, Fifth Edition, 1968, Howard W. Sams and Co., no ISBN, Library of Congress Card No. 43-14665 chapter 13
17. Arrillaga, Jos; Liu, Yonghe H; Watson, Neville R; Murray, Nicholas J. Self-Commutating Converters for High Power Applications. John Wiley & Sons. ISBN 978-0-470-68212-8.
18. Idaho National Laboratory (2007). "Harvesting the sun's energy with antennas". Retrieved 2008-10-03.

CHAPTER 8

Switched-Mode Power Supply

Chapter Contents

A switched-mode power supply (SMPS) is an electronic circuit that converts power using switching devices that are turned on and off at high frequencies, and storage components such as inductors or capacitors to supply power when the switching device is in its non-conduction state.

Switching power supplies have high efficiencies and are widely used in a variety of electronic equipment, including computers and other sensitive equipment requiring stable and efficient power supply.

A switched-mode power supply is also known as a switch-mode power supply or switching-mode power supply.

INTRODUCTION

A switched-mode power supply (switching-mode power supply, switch-mode power supply, switched power supply, SMPS, or switcher) is an electronic power supply that incorporates a switching regulator to convert electrical power efficiently. Like other power supplies, an SMPS transfers power from a DC or AC source (often mains power), to DC loads, such as a personal computer, while converting voltage and current characteristics. Unlike a linear power supply, the pass transistor of a switching-mode supply continually switches between low-dissipation, full-on and full-off states, and spends very little time in the high dissipation transitions, which minimizes wasted energy. Ideally, a switched-mode power supply dissipates no power. Voltage regulation is achieved by varying the ratio of on-to-off time. In contrast, a linear power supply regulates the output voltage by continually dissipating power in the pass transistor. This higher power conversion efficiency is an important advantage of a switched-mode power supply. Switched-mode power supplies may also be substantially smaller and lighter than a linear supply due to the smaller transformer size and weight.

Switching regulators are used as replacements for linear regulators when higher efficiency, smaller size or lighter weight are required. They are, however, more complicated; their switching currents can cause electrical noise problems if not carefully suppressed, and simple designs may have a poor power factor.

Interior view of an ATX SMPS: below

A: input EMI filtering and bridge rectifier;

B: input filter capacitors;

Between B and C: primary side heat sink;

C: transformer;

Between C and D: secondary side heat sink;

D: output filter coil;

E: output filter capacitors.

The coil and large yellow capacitor below E are additional input filtering components that are mounted directly on the power input connector and are not part of the main circuit board.

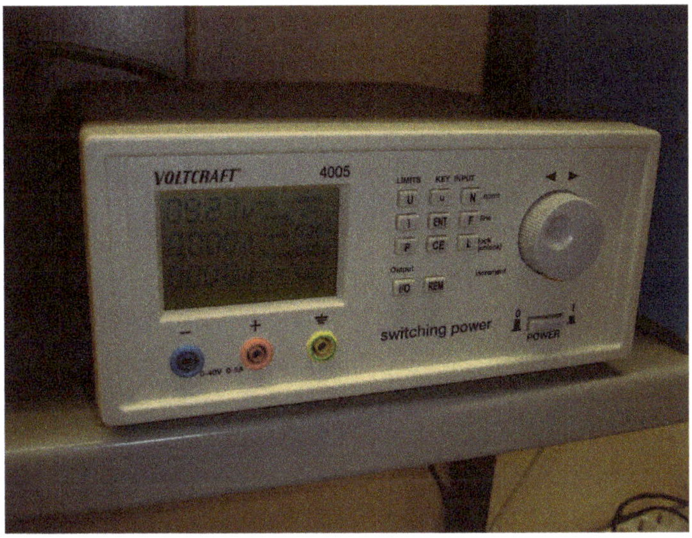

An adjustable switched-mode power supply for laboratory use.

HISTORY

1836

Induction coils use switches to generate high voltages.

1910

An inductive discharge ignition system invented by Charles F. Kettering and his company Dayton Engineering Laboratories Company (Delco) goes into production for Cadillac. The Kettering ignition system is a mechanically-switched version of a flyback boost converter; the transformer is the ignition coil. Variations of this ignition system were used in all non-diesel internal combustion engines until the 1960s when it was displaced with capacitive discharge ignition systems.

1926

On 23 June, British inventor Philip Ray Coursey applies for a patent in his country and United States, for his "Electrical Condenser". The patent mentions high frequency welding and furnaces, among other uses.

ca 1936

Car radios used electromechanical vibrators to transform the 6 V battery supply to a suitable B+ voltage for the vacuum tubes.

1959

Transistor oscillation and rectifying converter power supply system U.S. Patent 3,040,271 is filed by Joseph E. Murphy and Francis J. Starzec, from General Motors Company

1970

Tektronix starts using High-Efficiency Power Supply in its 7000-series oscilloscopes produced from about 1970 to 1995.

1972

HP-35, Hewlett-Packard's first pocket calculator, is introduced with transistor switching power supply for light-emitting diodes, clocks, timing, ROM, and registers.

1973

Xerox uses switching power supplies in the Alto minicomputer

1977

Apple II is designed with a switching mode power supply. "*Rod Holt was brought in as product engineer and there were several flaws in Apple II that were never publicized. One thing Holt has to his credit is that he created the switching power supply that allowed us to do a very lightweight computer*".

1980

The HP8662A 10 kHz – 1.28 GHz synthesized signal generator went with a switched mode power supply.

EXPLANATION

A linear regulator provides the desired output voltage by dissipating excess power in ohmic losses (e.g., in a resistor or in the collector–emitter region of a pass transistor in its active mode). A linear regulator regulates either output voltage or current by dissipating the excess electric power in the form of heat, and hence its maximum power efficiency is voltage-out/voltage-in since the volt difference is wasted.

In contrast, a switched-mode power supply regulates either output voltage or current by switching ideal storage elements, like inductors and capacitors, into and out of different electrical configurations. Ideal switching elements (e.g., transistors operated outside of their active mode) have no resistance when "closed" and carry no current when "open", and so the converters can theoretically operate with 100% efficiency (i.e., all input power is delivered to the load; no power is wasted as dissipated heat).

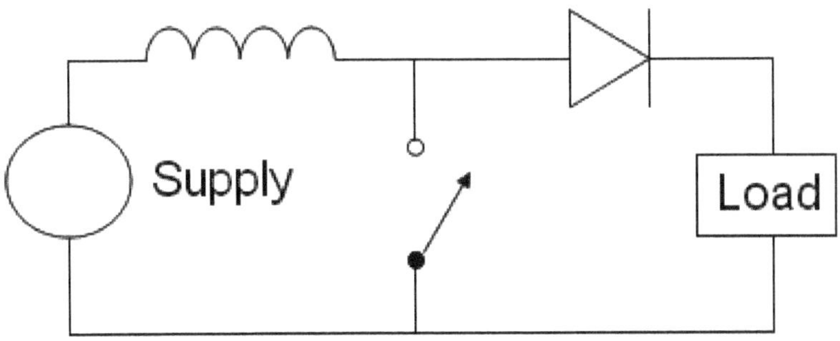

The basic schematic of a boost converter.

For example, if a DC source, an inductor, a switch, and the corresponding electrical ground are placed in series and the switch is driven by a square wave, the peak-to-peak voltage of the waveform measured across the switch can exceed the input voltage from the DC source. This is because the inductor responds to changes in current by inducing its own voltage to counter the change in current, and this voltage adds to the source voltage while the switch is open. If a diode-and-capacitor combination is placed in parallel to the switch, the peak voltage can be stored in the capacitor, and the capacitor can be used as a DC source with an output voltage greater than the DC voltage driving the circuit. This boost converter acts like a step-up transformer for DC signals. A buck–boost converter works in a similar manner, but yields an output voltage which is opposite in polarity to the input voltage. Other buck circuits exist to boost the average output current with a reduction of voltage.

In a SMPS, the output current flow depends on the input power signal, the storage elements and circuit topologies used, and also on the pattern used (e.g., pulse-width modulation with an adjustable duty cycle) to drive the switching elements. The spectral density of these switching waveforms has energy concentrated at relatively high frequencies. As such, switching transients and ripple introduced onto the output waveforms can be filtered with a small LC filter.

ADVANTAGES AND DISADVANTAGES

The main advantage of the switching power supply is greater efficiency than linear regulators because the switching transistor dissipates little power when acting as a switch.

Other advantages include smaller size and lighter weight from the elimination of heavy line-frequency transformers, and comparable heat generation. Standby power loss is often much less than transformers.

Disadvantages include greater complexity, the generation of high-amplitude, high-frequency energy that the low-pass filter must block to avoid electromagnetic interference (EMI), a ripple voltage at the switching frequency and the harmonic frequencies thereof.

Very low cost SMPSs may couple electrical switching noise back onto the mains power line, causing interference with A/V equipment connected to the same phase. Non-power-factor-corrected SMPSs also cause harmonic distortion.

SMPS AND LINEAR POWER SUPPLY COMPARISON

There are two main types of regulated power supplies available: SMPS and linear. The following table compares linear regulated and unregulated AC-to-DC supplies with switching regulators in general:

Comparison of a linear power supply and a switched-mode power supply			
	Linear power supply	**Switching power supply**	**Notes**
Size and weight	Heatsinks for high power linear regulators add size and weight. Transformers, if used, are large due to low operating frequency (mains power frequency is at 50 or 60 Hz); otherwise can be compact due to low component count.	Smaller transformer (if used; else inductor) due to higher operating frequency (typically 50 kHz – 1 MHz). Size and weight of adequate RF shielding may be significant.	A transformer's power handling capacity of given size and weight increases with frequency provided that hysteresis losses can be kept down. Therefore, higher operating frequency means either a higher capacity or smaller transformer.
Output voltage	With transformer used, any voltages available; if transformerless, limited to what can be achieved with a voltage doubler. If unregulated, voltage varies significantly with load.	Any voltages available, limited only by transistor breakdown voltages in many circuits. Voltage varies little with load.	A SMPS can usually cope with wider variation of input before the output voltage changes.
Efficiency, heat, and powerdissipation	If regulated: efficiency largely depends on voltage difference between input and output; output voltage is regulated by dissipating excess power as heat resulting in a typical efficiency of 30–40%. If unregulated, transformer iron and copper losses may be the only significant sources of inefficiency.	Output is regulated using duty cycle control; the transistors are switched fully on or fully off, so very little resistive losses between input and the load. The only heat generated is in the non-ideal aspects of the components and quiescent current in the control circuitry.	Switching losses in the transistors (especially in the short part of each cycle when the device is partially on), on-resistance of the switching transistors, equivalent series resistance in the inductor and capacitors, and core losses in the inductor, and rectifier voltage drop contribute to a typical efficiency of 60–70%. However, by optimizing SMPS design (such as choosing the optimal switching frequency, avoiding saturation of inductors, and active rectification), the amount of power loss and heat can be minimized; a good design can have an efficiency of 95%.

Complexity	Unregulated may be simply a diode and capacitor; regulated has a voltage-regulating circuit and a noise-filtering capacitor; usually a simpler circuit (and simpler feedback loop stability criteria) than switched-mode circuits.	Consists of a controller IC, one or several power transistors and diodes as well as a power transformer, inductors, and filter capacitors. Some design complexities present (reducing noise/interference; extra limitations on maximum ratings of transistors at high switching speeds) not found in linear regulator circuits.	In switched-mode mains (AC-to-DC) supplies, multiple voltages can be generated by one transformer core, but that can introduce design/use complications: for example it may place *minimum* output current restrictions on one output. For this SMPSs have to use duty cycle control. One of the outputs has to be chosen to feed the voltage regulation feedback loop (usually 3.3 V or 5 V loads are more fussy about their supply voltages than the 12 V loads, so this drives the decision as to which feeds the feedback loop. The other outputs usually track the regulated one pretty well). Both need a careful selection of their transformers. Due to the high operating frequencies in SMPSs, the stray inductance and capacitance of the printed circuit board traces become important.
Radio frequency interference	Mild high-frequency interference may be generated by AC rectifier diodes under heavy current loading, while most other supply types produce no high-frequency interference. Some mains hum induction into unshielded cables, problematical for low-signal audio.	EMI/RFI produced due to the current being switched on and off sharply. Therefore, EMI filters and RF shielding are needed to reduce the disruptive interference.	Long wires between the components may reduce the high frequency filter efficiency provided by the capacitors at the inlet and outlet. Stable switching frequency may be important.
Electronic noise at the output terminals	Unregulated PSUs may have a little AC ripple superimposed upon the DC component at twice mains frequency (100–120 Hz). It can cause audible mains hum in audio equipment, brightness ripples or banded distortions in analog security cameras.	Noisier due to the switching frequency of the SMPS. An unfiltered output may cause glitches in digital circuits or noise in audio circuits.	This can be suppressed with capacitors and other filtering circuitry in the output stage. With a switched mode PSU the switching frequency can be chosen to keep the noise out of the circuits working frequency band (e.g., for audio systems above the range of human hearing)
Electronic noise at the input terminals	Causes harmonic distortion to the input AC, but relatively little or no high frequency noise.	Very low cost SMPS may couple electrical switching noise back onto the mains power line, causing interference with A/V equipment connected to the same phase. Non power-factor-corrected SMPSs also cause harmonic distortion.	This can be prevented if a (properly earthed) EMI/RFI filter is connected between the input terminals and the bridge rectifier.
Acoustic noise	Faint, usually inaudible mains hum, usually due to vibration of windings in the transformer or magnetostriction.	Usually inaudible to most humans, unless they have a fan or are unloaded/malfunctioning, or use a switching frequency within the audio range, or the laminations of the coil vibrate at a subharmonic of the operating frequency.	The operating frequency of an unloaded SMPS is sometimes in the audible human range, and may sound subjectively quite loud for people whose hearing is very sensitive to the relevant frequency range.

Power factor	Low for a regulated supply because current is drawn from the mains at the peaks of the voltage sinusoid, unless a choke-input or resistor-input circuit follows the rectifier (now rare).	Ranging from very low to medium since a simple SMPS without PFC draws current spikes at the peaks of the AC sinusoid.	Active/passive power factor correction in the SMPS can offset this problem and are even required by some electric regulation authorities, particularly in the EU. The internal resistance of low-power transformers in linear power supplies usually limits the peak current each cycle and thus gives a better power factor than many switched-mode power supplies that directly rectify the mains with little series resistance.
Inrush current	Large current when mains-powered linear power supply equipment is switched on until magnetic flux of transformer stabilises and capacitors charge completely, unless a slow-start circuit is used.	Extremely large peak "in-rush" surge current limited only by the impedance of the input supply and any series resistance to the filter capacitors.	Empty filter capacitors initially draw large amounts of current as they charge up, with larger capacitors drawing larger amounts of peak current. Being many times above the normal operating current, this greatly stresses components subject to the surge, complicates fuse selection to avoid nuisance blowing and may cause problems with equipment employing overcurrent protection such as uninterruptible power supplies. Mitigated by use of a suitable soft-start circuit or series resistor.
Risk of electric shock	Supplies with transformers isolate the incoming power supply from the powered device and so allow metalwork of the enclosure to be grounded safely. Dangerous if primary/secondary insulation breaks down, unlikely with reasonable design. Transformerless mains-operated supply dangerous. In both linear and switch-mode the mains, and possibly the output voltages, are hazardous and must be well-isolated.	Common rail of equipment (including casing) is energized to half the mains voltage, but at high impedance, unless equipment is earthed/grounded or doesn't contain EMI/RFI filtering at the input terminals.	Due to regulations concerning EMI/RFI radiation, many SMPS contain EMI/RFI filtering at the input stage before the bridge rectifier consisting of capacitors and inductors. Two capacitors are connected in series with the Live and Neutral rails with the Earth connection in between the two capacitors. This forms a capacitive divider that energizes the common rail at half mains voltage. Its high impedance current source can provide a tingling or a 'bite' to the operator or can be exploited to light an Earth Fault LED. However, this current may cause nuisance tripping on the most sensitive residual-current devices.

Risk of equipment damage	Very low, unless a short occurs between the primary and secondary windings or the regulator fails by shorting internally.	Can fail so as to make output voltage very high. Stress on capacitors may cause them to explode. Can in some cases destroy input stages in amplifiers if floating voltage exceeds transistor base-emitter breakdown voltage, causing the transistor's gain to drop and noise levels to increase. Mitigated by good failsafe design. Failure of a component in the SMPS itself can cause further damage to other PSU components; can be difficult to troubleshoot.	The floating voltage is caused by capacitors bridging the primary and secondary sides of the power supply. Connection to earthed equipment will cause a momentary (and potentially destructive) spike in current at the connector as the voltage at the secondary side of the capacitor equalizes to earth potential.

THEORY OF OPERATION

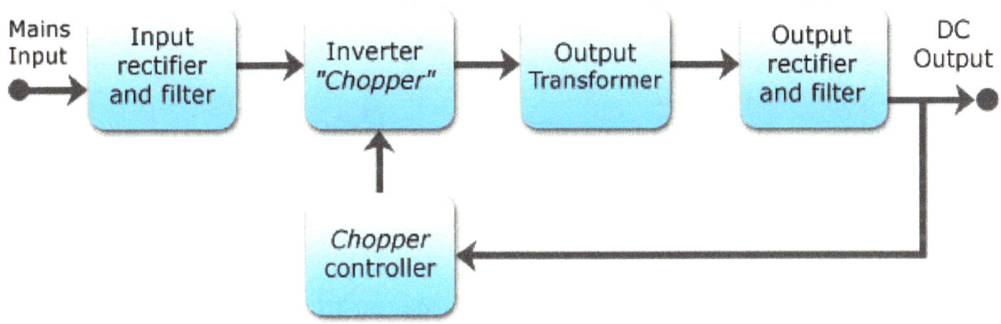

Block diagram of a mains operated AC/DC SMPS with output voltage regulation.

Input Rectifier Stage

If the SMPS has an AC input, then the first stage is to convert the input to DC. This is called *rectification*. A SMPS with a DC input does not require this stage. In some power supplies (mostly computer ATX power supplies), the rectifier circuit can be configured as a voltage doubler by the addition of a switch operated either manually or automatically. This feature permits operation from power sources that are normally at 115 V or at 230 V. The rectifier produces an unregulated DC voltage which is then sent to a large filter capacitor. The current drawn from the mains supply by this rectifier circuit occurs in short pulses around the AC voltage peaks. These pulses have significant high frequency energy which reduces the power factor. To correct for this, many newer SMPS will use a special PFC circuit to make the input current follow the sinusoidal shape of the AC input voltage, correcting the power factor. Power supplies that use Active PFC usually are auto-ranging, supporting input voltages from ~100 VAC – 250 VAC, with no input voltage selector switch.

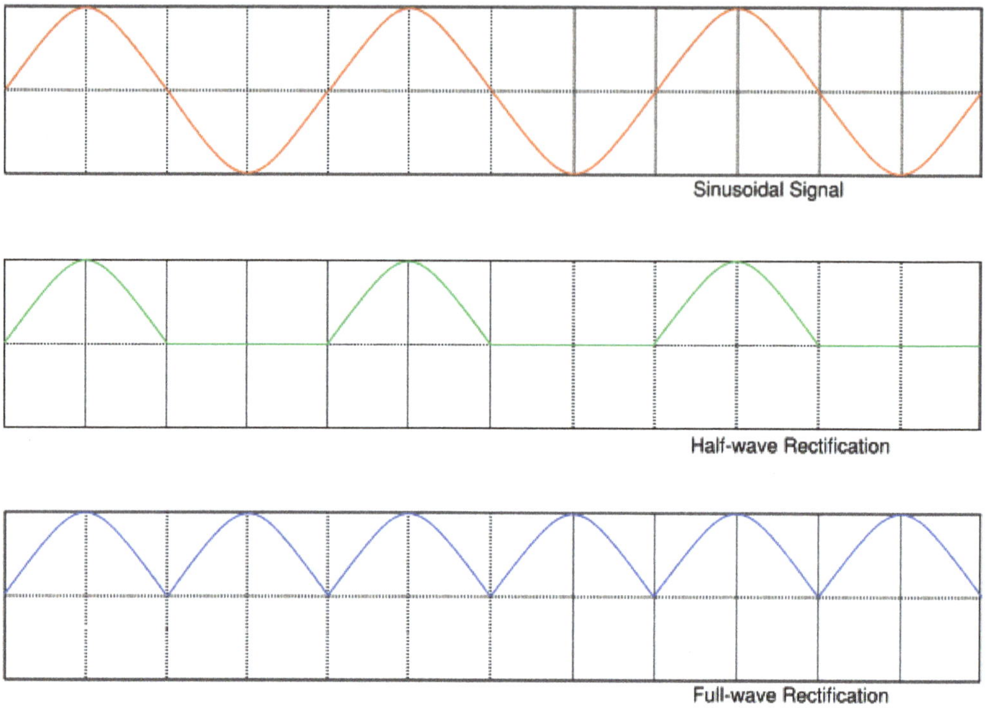

AC, half-wave and full-wave rectified signals.

An SMPS designed for AC input can usually be run from a DC supply, because the DC would pass through the rectifier unchanged. If the power supply is designed for 115 VAC and has no voltage selector switch, the required DC voltage would be 163 VDC (115 × √2). This type of use may be harmful to the rectifier stage, however, as it will only use half of diodes in the rectifier for the full load. This could possibly result in overheating of these components, causing them to fail prematurely. On the other hand, if the power supply has a voltage selector switch, based on the Delon circuit, for 115/230V (computer ATX power supplies typically are in this category), the selector switch would have to be put in the 230 V position, and the required voltage would be 325 VDC (230 × √2). The diodes in this type of power supply will handle the DC current just fine because they are rated to handle double the nominal input current when operated in the 115 V mode, due to the operation of the voltage doubler. This is because the doubler, when in operation, uses only half of the bridge rectifier and runs twice as much current through it. It is uncertain how an Auto-ranging/Active-PFC type power supply would react to being powered by DC.

Inverter Stage

This section refers to the block marked *chopper* in the diagram.

The inverter stage converts DC, whether directly from the input or from the rectifier stage described above, to AC by running it through a power oscillator, whose output transformer is very small with few windings at a frequency of tens or hundreds of kilohertz. The frequency is usually chosen to be above 20 kHz, to make it inaudible to humans. The switching is implemented as a multistage (to achieve high gain) MOSFET amplifier. MOSFETs are a type of transistor with a low on-resistance and a high current-handling capacity.

Voltage Converter and Output Rectifier

If the output is required to be isolated from the input, as is usually the case in mains power supplies, the inverted AC is used to drive the primary winding of a high-frequency transformer. This converts the voltage up or down to the required output level on its secondary winding. The output transformer in the block diagram serves this purpose.

If a **DC** output is required, the **AC** output from the transformer is rectified. For output voltages above ten volts or so, ordinary silicon diodes are commonly used. For lower voltages, Schottky diodes are commonly used as the rectifier elements; they have the advantages of faster recovery times than silicon diodes (allowing low-loss operation at higher frequencies) and a lower voltage drop when conducting. For even lower output voltages, MOSFETs may be used as synchronous rectifiers; compared to Schottky diodes, these have even lower conducting state voltage drops.

The rectified output is then smoothed by a filter consisting of inductors and capacitors. For higher switching frequencies, components with lower capacitance and inductance are needed.

Simpler, non-isolated power supplies contain an inductor instead of a transformer. This type includes *boost converters*, *buck converters*, and the *buck-boost converters*. These belong to the simplest class of single input, single output converters which use one inductor and one active switch. The buck converter reduces the input voltage in direct proportion to the ratio of conductive time to the total switching period, called the duty cycle. For example an ideal buck converter with a 10 V input operating at a 50% duty cycle will produce an average output voltage of 5 V. A feedback control loop is employed to regulate the output voltage by varying the duty cycle to compensate for variations in input voltage. The output voltage of a boost converter is always greater than the input voltage and the buck-boost output voltage is inverted but can be greater than, equal to, or less than the magnitude of its input voltage. There are many variations and extensions to this class of converters but these three form the basis of almost all isolated and non-isolated DC to DC converters. By adding a second inductor the Ćuk and SEPIC converters can be implemented, or, by adding additional active switches, various bridge converters can be realized.

Other types of SMPSs use a capacitor-diode voltage multiplier instead of inductors and transformers. These are mostly used for generating high voltages at low currents (*Cockcroft-Walton generator*). The low voltage variant is called charge pump.

Regulation

This charger for a small device such as a mobile phone is a simple off-line switching power supply with a European plug.

A feedback circuit monitors the output voltage and compares it with a reference voltage, as shown in the block diagram above. Depending on design and safety requirements, the controller may contain an isolation mechanism (such as an opto-coupler) to isolate it from the DC output. Switching supplies in computers, TVs and VCRs have these opto-couplers to tightly control the output voltage.

Open-loop regulators do not have a feedback circuit. Instead, they rely on feeding a constant voltage to the input of the transformer or inductor, and assume that the output will be correct. Regulated designs compensate for the impedance of the transformer or coil. Monopolar designs also compensate for the magnetic hysteresis of the core.

The feedback circuit needs power to run before it can generate power, so an additional non-switching power-supply for stand-by is added.

TRANSFORMER DESIGN

Any switched-mode power supply that gets its power from an AC power line (called an "off-line" converter) requires a transformer for galvanic isolation. Some DC-to-DC converters may also include a transformer, although isolation may not be critical in these cases. SMPS transformers run at high frequency. Most of the cost savings (and space savings) in off-line power supplies result from the smaller size of the high frequency transformer compared to the 50/60 Hz transformers formerly used. There are additional design tradeoffs.

The terminal voltage of a transformer is proportional to the product of the core area, magnetic flux, and frequency. By using a much higher frequency, the core area (and so the mass of the core) can be greatly reduced. However, core losses increase at higher frequencies. Cores generally use ferrite material which has a low loss at the high frequencies and high flux densities used. The laminated iron cores of lower-frequency (<400 Hz) transformers would be unacceptably lossy at switching frequencies of a few kilohertz. Also, more energy is lost during transitions of the switching semiconductor at higher frequencies. Furthermore, more attention to the physical layout of the circuit board is required as parasitics become more significant, and the amount of electromagnetic interference will be more pronounced.

Copper Loss

At low frequencies (such as the line frequency of 50 or 60 Hz), designers can usually ignore the skin effect. For these frequencies, the skin effect is only significant when the conductors are large, more than 0.3 inches (7.6 mm) in diameter.

Switching power supplies must pay more attention to the skin effect because it is a source of power loss. At 500 kHz, the skin depth in copper is about 0.003 inches (0.076 mm) – a dimension smaller than the typical wires used in a power supply. The effective resistance of conductors increases, because current concentrates near the surface of the conductor and the inner portion carries less current than at low frequencies.

The skin effect is exacerbated by the harmonics present in the high speed PWM switching waveforms. The appropriate skin depth is not just the depth at the fundamental, but also the skin depths at the harmonics.

In addition to the skin effect, there is also a proximity effect, which is another source of power loss.

POWER FACTOR

Simple off-line switched mode power supplies incorporate a simple full-wave rectifier connected to a large energy storing capacitor. Such SMPSs draw current from the AC line in short pulses when the mains instantaneous voltage exceeds the voltage across this capacitor. During the remaining portion of the AC cycle the capacitor provides energy to the power supply.

As a result, the input current of such basic switched mode power supplies has high harmonic content and relatively low power factor. This creates extra load on utility lines, increases heating of building wiring, the utility transformers, and standard AC electric motors, and may cause stability problems in some applications such as in emergency generator systems or aircraft generators. Harmonics can be removed by filtering, but the filters are expensive. Unlike displacement power factor created by linear inductive or capacitive loads, this distortion cannot be corrected by addition of a single linear component. Additional circuits are required to counteract the effect of the brief current pulses. Putting a current regulated boost chopper stage after the off-line rectifier (to charge the storage capacitor) can correct the power factor, but increases the complexity and cost.

In 2001, the European Union put into effect the standard IEC/EN61000-3-2 to set limits on the harmonics of the AC input current up to the 40th harmonic for equipment above 75 W. The standard defines four classes of equipment depending on its type and current waveform. The most rigorous limits (class D) are established for personal computers, computer monitors, and TV receivers. To comply with these requirements, modern switched-mode power supplies normally include an additional power factor correction (PFC) stage.

TYPES

Switched-mode power supplies can be classified according to the circuit topology. The most important distinction is between isolated converters and non-isolated ones.

Non-Isolated Topologies

Non-isolated converters are simplest, with the three basic types using a single inductor for energy storage. In the voltage relation column, *D* is the duty cycle of the converter, and can vary from 0 to 1. The input voltage (V_1) is assumed to be greater than zero; if it is negative, for consistency, negate the output voltage (V_2).

Type	Typical Power [W]	Relative cost	Energy storage	Voltage relation	Features
Buck	0–1,000	1.0	Single inductor	$0 \leq$ Out $\leq$ In, $V_2 = DV_1$	Current is continuous at output.
Boost	0–5,000	1.0	Single inductor	Out $\geq$ In, $V_2 = \frac{1}{1-D} V_1$	Current is continuous at input.
Buck-boost	0–150	1.0	Single inductor	Out ≤ 0, $V_2 = -\frac{D}{1-D} V_1$	Current is dis-continuous at both input and output.
Split-pi (or, boost-buck)	0–4,500	>2.0	Two inductors and three capacitors	Up or down	Bidirectional power control; in or out
Ćuk			Capacitor and two inductors	Any inverted, $V_2 = -\frac{D}{1-D} V_1$	Current is continuous at input *and* output
SEPIC			Capacitor and two inductors	Any, $V_2 = \frac{D}{1-D} V_1$	Current is continuous at input
Zeta			Capacitor and two inductors	Any, $V_2 = \frac{D}{1-D} V_1$	Current is continuous at output
Charge pump / Switched capacitor			Capacitors only		No magnetic energy storage is needed to achieve conversion, however high efficiency power processing is normally limited to a discrete set of conversion ratios.

When equipment is human-accessible, voltage and power limits of <=42.4 V peak/60 V dc and 250 VA apply for Safety Certification (UL, CSA, VDE approval).

The buck, boost, and buck-boost topologies are all strongly related. Input, output and ground come together at one point. One of the three passes through an inductor on the way, while the other two pass through switches. One of the two switches must be active (e.g., a transistor), while the other can be a diode. Sometimes, the topology can be changed simply by re-labeling the connections. A 12 V input, 5 V output buck converter can be converted to a 7 V input, −5 V output buck-boost by grounding the *output* and taking the output from the *ground* pin.

Likewise, SEPIC and Zeta converters are both minor rearrangements of the Ćuk converter.

The Neutral Point Clamped (NPC) topology is used in power supplies and active filters and is mentioned here for completeness.

Switchers become less efficient as duty cycles become extremely short. For large voltage changes, a transformer (isolated) topology may be better.

Isolated Topologies

All isolated topologies include a transformer, and thus can produce an output of higher or lower voltage than the input by adjusting the turns ratio. For some topologies, multiple windings can be placed on the transformer to produce multiple output voltages. Some converters use the transformer for energy storage, while others use a separate inductor.

Type	Power [W]	Relative cost	Input range [V]	Energy storage	Features
Flyback	0–250	1.0	5–600	Mutual Inductors	Isolated form of the buck-boost converter.[1]
Ringing choke converter (RCC)	0–150	1.0	5–600	Transformer	Low-cost self-oscillating flyback variant.
Half-forward	0–250	1.2	5–500	Inductor	
Forward[2]	100-200		60–200	Inductor	Isolated form of buck converter
Resonant forward	0–60	1.0	60–400	Inductor and capacitor	Single rail input, unregulated output, high efficiency, low EMI.
Push-pull	100–1,000	1.75	50–1,000	Inductor	
Half-bridge	0–2,000	1.9	50–1,000	Inductor	
Full-bridge	400–5,000	>2.0	50–1,000	Inductor	Very efficient use of transformer, used for highest powers.
Resonant, zero voltage switched	>1,000	>2.0		Inductor and capacitor	
Isolated Ćuk				Two capacitors and two inductors	

Zero voltage switched mode power supplies require only small heatsinks as little energy is lost as heat. This allows them to be small. This ZVS can deliver more than 1 kilowatt. Transformer is not shown.

- **1** Flyback converter logarithmic control loop behavior might be harder to control than other types.
- **2** The forward converter has several variants, varying in how the transformer is "reset" to zero magnetic flux every cycle.

Quasi-Resonant Zero-Current/Zero-Voltage Switch

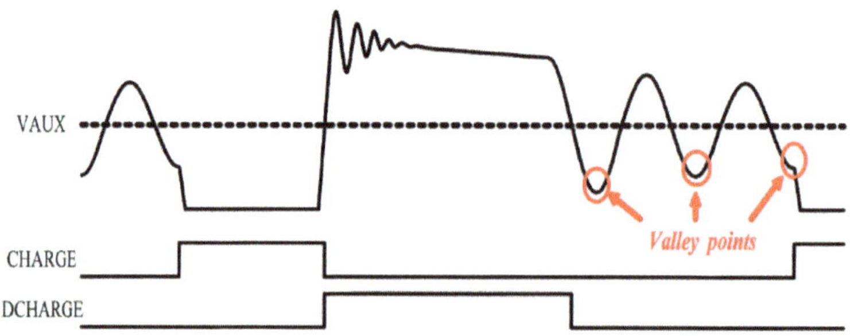

Quasi-resonant switching switches when the voltage is at a minimum and a valley is detected

In a quasi-resonant zero-current/zero-voltage switch (ZCS/ZVS) "each switch cycle delivers a quantized 'packet' of energy to the converter output, and switch turn-on and turn-off occurs at zero current and voltage, resulting in an essentially lossless switch." Quasi-resonant switching, also known as *valley switching*, reduces EMI in the power supply by two methods:

- By switching the bipolar switch when the voltage is at a minimum (in the valley) to minimize the hard switching effect that causes EMI.
- By switching when a valley is detected, rather than at a fixed frequency, introduces a natural frequency jitter that spreads the RF emissions spectrum and reduces overall EMI.

EFFICIENCY AND EMI

Higher input voltage and synchronous rectification mode makes the conversion process more efficient. The power consumption of the controller also has to be taken into account. Higher switching frequency allows component sizes to be shrunk, but can produce more RFI. A resonant forward converter produces the lowest EMI of any SMPS approach because it uses a soft-switching resonant waveform compared with conventional hard switching.

Failure Modes

Power supplies which use capacitors suffering from the capacitor plague may experience premature failure when the capacitance drops to 4% of the original value. This usually causes the switching semiconductor to fail in a conductive way. That may expose connected loads to the full input volt and current, and precipitate wild oscillations in output.

Failure of the switching transistor is common. Due to the large switching voltages this transistor must handle (around 325 V for a 230 V_{AC} mains supply), these transistors often short out, in turn immediately blowing the main internal power fuse.

Precautions

The main filter capacitor will often store up to 325 volts long after the power cord has been removed from the wall. Not all power supplies contain a small "bleeder" resistor to slowly discharge this capacitor. Any contact with this capacitor may result in a severe electrical shock.

The primary and secondary sides may be connected with a capacitor to reduce EMI and compensate for various capacitive couplings in the converter circuit, where the transformer is one. This may result in electric shock in some cases. The current flowing from line or neutral through a 2 kΩ resistor to any accessible part must, according to IEC 60950, be less than 250 μA for IT equipment.

APPLICATIONS

Switched-mode power supply units (PSUs) in domestic products such as personal computers often have universal inputs, meaning that they can accept power from mains supplies throughout the world, although a manual voltage range switch may be required. Switch-mode power supplies can tolerate a wide range of power frequencies and voltages.

Switched mode mobile phone charger.

A 450 Watt SMPS for use in personal computers with the power input, fan, and output cords visible.

Due to their high volumes mobile phone chargers have always been particularly cost sensitive. The first chargers were linear power supplies but they quickly moved to the cost effective ringing choke converter (RCC) SMPS topology, when new levels of efficiency were required. Recently, the demand for even lower no-load power requirements in the application has meant that flyback topology is being used more widely; primary side sensing flyback controllers are also helping to cut the bill of materials (BOM) by removing secondary-side sensing components such as optocouplers.

Switched-mode power supplies are used for DC to DC conversion as well. In automobiles where heavy vehicles use a nominal 24 V_{DC} cranking supply, 12V for accessories may be furnished through a DC/DC switch-mode supply. This has the advantage over tapping the battery at the 12V position (using half the cells) that all the 12V load is evenly divided over all cells of the 24V battery. In industrial settings such as telecommunications racks, bulk power may be distributed at a low DC voltage (from a battery back up system, for example) and individual equipment items will have DC/DC switched-mode converters to supply whatever voltages are needed.

REFERENCES

1. US 1037492, Kettering, Charles F., "Ignition system", published 2 November 1910, issued 3 September 1912
2. US 1754265, Coursey, Philip Ray, "Electrical Condenser", published 23 June 1926, issued 15 April 1930
3. "When was the SMPS power supply invented?". electronicspoint.com.
4. "Electrical condensers (Open Library)". openlibrary.org.
5. Cadillac model 5-X, a 5 tube supherheterodyne radio, used a synchronous vibrator to generate its B+ supply. RadioMuseum.org, http://www.radiomuseum.org/r/cadillacge_5x.html#a
6. google.com/patents - Transistor converter power supply system
7. slack.com - Test Equipment and Electronics Information
8. kahrs.us - The Tektronix 7000 Oscilloscope series
9. tek.com - 7000 Series oscilloscopes FAQ
10. docmesure.free.fr - TEKSCOPE March 1971 7704 High-Efficiency Power Supply (service manual march-1971 .pdf)
11. jacques-laporte.org - The HP-35's Power unit and other vintage HP calculators.
12. Y Combinator's Xerox Alto: restoring the legendary 1970s GUI computer
13. businessinsider.com - EXCLUSIVE: Interview With Apple's First CEO Michael Scott 2011-05-24
14. "HP 3048A". hpmemoryproject.org.
15. "Energy Savings Opportunity by Increasing Power Supply Efficiency".
16. "Ban Looms for External Transformers". 080224 sound.whsites.net
17. "DC Power Production, Delivery and Utilization, An EPRI White Paper" (PDF). Page 9 080317 mydocs.epri.com
18. Notes on the Troubleshooting and Repair of Small Switchmode Power Supplies: Switching between 115 VAC and 230 VAC input. Search the page for "doubler" for more info. Retrieved March 2013.

19. Foutz, Jerrold. "Switching-Mode Power Supply Design Tutorial Introduction". Retrieved 2008-10-06.
20. ON Semiconductor (July 11, 2002). "SWITCHMODE Power Supplies—Reference Manual and Design Guide" (PDF). Retrieved 2011-11-17.
21. "An active power filter implemented with multilevel single-phase NPC converters". 2011. Retrieved 2013-03-15.
22. "DC-DC Converter Basics". 090112 powerdesigners.com
23. "DC-DC CONVERTERS: A PRIMER" (PDF). 090112 jaycar.com.au Page 4
24. "Heinz Schmidt-Walter". h-da.de.
25. Irving, Brian T.; Jovanović, Milan M. (March 2002), Analysis and Design of Self-Oscillating Flyback Converter (PDF), Proc. IEEE Applied Power Electronics Conf. (APEC), pp. 897–903, retrieved 2009-09-30
26. "RDFC topology for linear replacement". 090725 camsemi.com Further information on resonant forward topology for consumer applications
27. "Gain Equalization Improves Flyback Performance Page of". 100517 powerelectronics.com
28. "Error - EDN". EDN.
29. "Bad Capacitors: Information and symptoms". 100211 lowyat.net
30. Electrical Power Quality and Utilization, Journal Vol. XV, No. 2, 2009: Estimation of Optimum Value of Y-Capacitor for Reducing Emi in Switch Mode Power Supplies

Index